Janko Petereit

Adaptive State × Time Lattices

A Contribution to Mobile Robot Motion Planning in Unstructured Dynamic Environments

Karlsruher Schriften zur Anthropomatik
Band 27
Herausgeber: Prof. Dr.-Ing. Jürgen Beyerer

Eine Übersicht aller bisher in dieser Schriftenreihe erschienenen Bände finden Sie am Ende des Buchs.

Adaptive State × Time Lattices

A Contribution to Mobile Robot Motion Planning in Unstructured Dynamic Environments

by
Janko Petereit

KIT Scientific Publishing

Dissertation, Karlsruher Institut für Technologie (KIT)
Fakultät für Informatik, 2016

Impressum

KIT Scientific Publishing

Karlsruher Institut für Technologie (KIT)
KIT Scientific Publishing
Straße am Forum 2
D-76131 Karlsruhe

KIT Scientific Publishing is a registered trademark of Karlsruhe Institute of Technology. Reprint using the book cover is not allowed.

www.ksp.kit.edu

Print on Demand 2017

ISSN 1863-6489
ISBN 978-3-7315-0580-8
DOI 10.5445/KSP/1000058693

Adaptive State × Time Lattices: A Contribution to Mobile Robot Motion Planning in Unstructured Dynamic Environments

zur Erlangung des akademischen Grades eines

Doktors der Ingenieurwissenschaften

von der KIT-Fakultät für Informatik
des Karlsruher Instituts für Technologie (KIT)

genehmigte

Dissertation

von

Janko Petereit

aus Kassel

Tag der mündlichen Prüfung: 22. Juli 2016
Erster Gutachter: Prof. Dr.-Ing. habil. Jürgen Beyerer
Zweiter Gutachter: Prof. Dr.-Ing. habil. Christoph Ament

Abstract

Besides comprehensive perception of the environment, planning the future motion of the robot is a key element for autonomous driving. Depending on the intended application, the requirements imposed on motion planning may vary. The characteristics of the environment, the potential presence of further participants in the scene, as well as kinematical and dynamical properties of the robotic platform significantly affect the complexity of motion planning. This thesis focuses on unmanned ground vehicles (UGVs) which operate in unstructured environments as they may occur, for example, after a natural disaster has struck. The consideration of dynamic obstacles (e.g., rescue workers or other rescue vehicles) requires to incorporate the dynamics of the robot itself into the planning. Since the environment may not be known in advance and may change at any time, a fast reaction is necessary.

Conventional motion planning for mobile robots usually consists of two separate steps: first, a path planning algorithm computes a global path to the goal on the basis of a map; afterward, a subsequent obstacle avoidance algorithm ensures that no collision with previously unknown or dynamic obstacles occurs. This subordinate obstacle avoidance algorithm usually operates only locally, which means that the global goal is not considered while planning the avoidance maneuver. As a result, the overall solution may be suboptimal in a global sense caused by the decoupling of the planning stages.

This thesis presents a holistic approach which drops the decoupling of global path planning and local obstacle avoidance and performs only *one* single integrated motion planning. The novel planning algorithm relies on the fact that the individual state variables are not equally relevant during the course of the future route to the goal. Instead, the relevance of the state variables depends on the (spatial and temporal) distance from the robot. For example, the robot's dynamics (i.e., velocity, time-parametrization) need to be considered in the vicinity of the robot and for the immediate future, to be able to react to dynamic obstacles; with increasing distance (and time, respectively) the dynamics of the robot become less important and it is sufficient to plan an ordinary, kinematically feasible path, which respects nonholonomic constraints. Restricting the high-dimensional planning to relevant regions considerably reduces the size of the search space and thus enables a faster generation of motion plans. In addition, also the requirements on planning resolution—and thus on the discretization of the planning result—decrease with increasing distance from the robot and from obstacles. Therefore, the ability to perform multi-resolution planning has also been integrated into the proposed planning framework.

The developed algorithm is based on *state* × *time lattices* which can be thought of as a generalized grid. The lattice points are connected by so-called motion primitives. In this thesis, a generic algorithm is proposed which is able to probabilistically generate motion primitives from arbitrary system models by running simulations of the robot's motion for random inputs. This is an offline procedure during the preparatory phase. In the subsequent planning phase, those motion primitives are used to construct a search graph that encodes the permissible state transitions. In this way, the motion planning problem is converted to the standard problem of finding the shortest path in a graph. The developed algorithm is designed to be universally usable and open for straightforward extension with additional capabilities. For example, the algorithm has been extended to enable combined motion planning for multiple waypoints, which makes the robot arrive at the next waypoint with an orientation that is optimal for the further travel to subsequent waypoints.

The result of the motion planning algorithm is of a hybrid nature: the first section is a real trajectory, which also includes temporal information; the next section is a dynamically feasible path, which at least contains the velocity information; the remaining section to the goal is represented by a pure kinematically feasible path, which only encodes the robot's position and orientation.

The proposed algorithm has been implemented in C++ and was evaluated using real recorded map data. It enables fast and reliable motion planning with a practical planning frequency of at least 10 Hz. The holistic approach, which integrates successive dimensionality reduction and the utilization of multiple resolutions, makes it possible to plan local maneuvers (and thus the imminent motion of the robot) which not only assess the global goal heuristically but rather consider it explicitly. This results in a globally optimal motion plan (of course, within the limits of the employed discretization).

Kurzfassung

Neben einer umfassenden Wahrnehmung der Umgebung ist das Planen der zukünftigen Roboterbewegung ein wichtiger Bestandteil des autonomen Fahrens. Je nach Anwendungsszenario werden hierbei unterschiedliche Anforderungen an die Bewegungsplanung gestellt. Die Beschaffenheit der Umgebung, die etwaige Gegenwart weiterer Akteure sowie die kinematischen und dynamischen Eigenschaften der Plattform beeinflussen maßgeblich die Komplexität der Bewegungsplanung. Der Fokus der vorliegenden Arbeit liegt auf Landrobotern, welche in einer unstrukturierten Umgebung, wie sie z. B. nach einer Naturkatastrophe vorliegt, agieren sollen. Die Einbeziehung dynamischer Hindernisse (z. B. Rettungshelfer oder andere Rettungsfahrzeuge) macht es erforderlich, auch die dynamischen Eigenschaften des Roboters selbst zu berücksichtigen. Da die Umgebung nicht a priori bekannt ist und sich auch während der Fahrt jederzeit verändern kann, ist eine kurze Reaktionszeit erforderlich.

Die herkömmliche Bewegungsplanung für mobile Roboter erfolgt meist zweistufig: Ein Pfadplanungsalgorithmus bestimmt auf Basis einer Karte einen globalen Pfad zum Ziel; anschließend stellt ein lokaler Hindernisvermeidungsalgorithmus sicher, dass keine Kollision mit vorher unbekannten oder dynamischen Objekten stattfindet. Dieser unterlagerte Hindernisvermeidungsalgorithmus operiert in der Regel nur lokal, was bedeutet, dass er bei der Planung eines Ausweichmanövers das globale Ziel nicht berücksichtigt. Die Folge ist ein – global betrachtet – suboptimales Ergebnis, welches aus der Separierung dieser beiden Planungsalgorithmen resultiert.

Diese Arbeit stellt ein ganzheitliches Verfahren vor, welches die Entkopplung von globaler Pfadplanung und lokaler Hindernisvermeidung aufhebt und stattdessen nur noch *eine* integrierte Bewegungsplanung vornimmt. Das neue Konzept nutzt hierbei die Tatsache aus, dass nicht alle Zustandsgrößen gleich relevant entlang des zukünftigen Weges bis zum Ziel sind. Vielmehr hängt die Relevanz der einzelnen Zustandsgrößen von der (räumlichen und zeitlichen) Entfernung zur aktuellen Roboterposition ab. So muss z. B. die Dynamik der Bewegung (Geschwindigkeit, zeitliche Parametrisierung) im Nahbereich und in der nahen Zukunft zwingend bei der Planung berücksichtigt werden, um korrekt auf dynamische Hindernisse reagieren zu können; mit steigendem Abstand (bzw. fortschreitender Zeit) ist die Dynamik aber immer weniger von Bedeutung und es ist ausreichend, einen gewöhnlichen, kinematisch zulässigen Pfad (unter Berücksichtigung nichtholonomer Randbedingungen) zu planen. Die Beschränkung der hochdimensionalen Planung auf die relevanten Bereiche führt zu einem reduzierten Suchraum und ermöglicht somit eine deutlich schnellere Bewegungsplanung. Ebenso sinken die Anforderungen an die räumliche Auflösung des Suchraums, und somit die Diskretisierung des Planungsergebnisses, mit zunehmender Entfernung von der Roboterposition und etwaigen Hindernissen. Daher wurde das Verfahren um entsprechende Fähigkeiten zur *Multi-Resolution*-Planung erweitert.

Der entwickelte Algorithmus basiert auf *state* × *time lattices*, welche als ein generalisiertes Gitter aufgefasst werden können. Die einzelnen Punkte des Gitters sind durch sogenannte Bewegungsprimitive miteinander verbunden. In dieser Arbeit wird ein generischer Algorithmus vorgestellt, welcher in der Lage ist, auf probabilistische Weise Bewegungsprimitive aus einem beliebigen Systemmodell zu generieren, indem die Bewegung des Roboters für zufällige Eingangsgrößen simuliert wird. Dies geschieht offline in einer Vorbereitungsphase. Anschließend werden die Bewegungsprimitive genutzt, um für die letztendliche Bewegungsplanung einen Graphen aufzuspannen, der die erlaubten Übergänge zwischen den Zuständen abbildet. Auf diese Art und Weise wird das Bewegungsplanungsproblem auf das Standardproblem der Suche eines kürzesten Weges in einem Graphen zurückgeführt. Der entwickelte Algorithmus ist sehr universell und offen gestaltet, sodass er sich leicht um weitere Fähigkeiten ergänzen lässt. Er wurde z. B. dahingehend erweitert, dass eine kombinierte Bewegungsplanung über mehrere Wegpunkte hinweg erfolgen kann, was den Roboter in die Lage versetzt, den jeweils nächsten Wegpunkt mit einer Ausrichtung zu erreichen, welche optimal für die weitere Fahrt zum darauffolgenden Wegpunkt ist.

Das Ergebnis der Bewegungsplanung ist hybrider Natur: Es besteht zum Teil aus einer echten Trajektorie, welche also auch eine zeitliche Information beinhaltet; daran schließt sich ein dynamisch zulässiger Pfad an, welcher zumindest noch eine Geschwindigkeitsinformation besitzt; der restliche Abschnitt bis zum Ziel wird durch einen nur noch kinematisch zulässigen Pfad repräsentiert, der also nur noch eine Information über Position und Ausrichtung des Roboters beinhaltet.

Der vorgestellte Algorithmus wurde vollständig mittels C++ umgesetzt und auf der Basis von realen Kartendaten evaluiert. Das Verfahren erlaubt eine schnelle, robuste Bewegungsplanung mit einer praxistauglichen Planungsfrequenz von mindestens 10 Hz. Durch die ganzheitliche Integration der schrittweisen Dimensionsreduktion und der Verwendung mehrerer Auflösungen wird es möglich, lokale Fahrmanöver (und somit die unmittelbare Roboterbewegung) zu planen, bei denen das globale Ziel nicht nur abgeschätzt, sondern vielmehr explizit berücksichtigt wird. Die Konsequenz ist ein global optimales Planungsergebnis (natürlich im Rahmen der gewählten Diskretisierung).

Acknowledgments

First and foremost, I would like to express my particular gratitude to my advisor Prof. Dr.-Ing. habil. Jürgen Beyerer for his continuous support and confidence in my work and for giving me the opportunity to pursue my PhD in a most encouraging research environment at the Fraunhofer Institute of Optronics, System Technologies and Image Exploitation IOSB. I would also like to thank my co-advisor Prof. Dr.-Ing. habil. Christoph Ament for his dedication and genuine interest in my research.

I am deeply grateful to my wonderful colleagues in the research group *multi-sensor systems*. This dissertation would not have been possible without their support. I would like to thank Christian Frey for always finding a way to provide us with the latest and greatest gadgets and for giving me the freedom to pursue my own ideas. I am most grateful for the support of Thomas Emter, Angelika Zube, and Christian Frese, who put tremendous effort into the development of the autonomy toolkit for our mobile robots. Special thanks go to my students for their unwavering dedication and commitment to many projects that substantially advanced the capabilities of our robots. I would like to thank Philipp Woock for inspiring debates on typography and Terry Atkinson for fruitful discussions about the peculiarities of the English language.

Last but not least, I would like to acknowledge the unbridled support and patience provided by my family and friends during this journey.

Contents

List of Figures

Acronyms

ARA*	Anytime Repairing A*
BVP	boundary value problem
CSI	Cauchy-Schwarz inequality
CPU	central processing unit
DAG	directed acyclic graph
DARPA	Defense Advanced Research Projects Agency
DWA	Dynamic Window Approach
EDT	Euclidean distance transform
GPS	Global Positioning System
HLUT	heuristic look-up table
IMU	inertial measurement unit
IP	interior point
LIDAR	light detection and ranging
NLP	nonlinear programming
NMPC	nonlinear model predictive control
OCP	optimal control problem
OMPL	Open Motion Planning Library
POMDP	Partially Observable Markov Decision Process
PRM	Probabilistic Roadmap
R&D	research and development

ROS	Robot Operating System
RRT	Rapidly-exploring Random Tree
SBPL	Search-Based Planning Library
SENEKA	Sensornetzwerk mit mobilen Robotern für das Katastrophenmanagement
SQP	sequential quadratic programming
UGV	unmanned ground vehicle

Symbols

Operators

$:=$	Definition
$\cup$	Union
$\cap$	Intersection
$\setminus$	Relative complement
$\times$	Cartesian product
$\subset$	Proper subset
$\subseteq$	Subset or equality
$\lvert\cdot\rvert$	Absolute value, cardinality of a set
$\lVert\cdot\rVert$	Euclidean norm
$\lceil\cdot\rceil$	Ceiling function
$\lfloor\cdot\rceil$	Round to nearest integer

Roman letters

a	Acceleration (input)
$\mathbf{b}$	Basis vector
c	Cost
C	Compatibility relation
d	Dimensionality level
d_{dyn}	Relative distance of robot and dynamic obstacle

d_{stat}	Distance to static obstacle
e_{q}^{r}	Quantization error for resolution level r
f	Priority queue key
$\mathbf{f}$	System model
g	Goal
g	Accumulated costs in graph search
$\underline{g}$	Projected accumulated costs in graph search
G	Set of goals
h	Heuristic cost estimate
J	Cost functional
J_{q}	Quantization loss
$l(m)$	Length of a motion primitive's trajectory
$\bar{l}(\tilde{\mathbf{s}})$	Length of trajectory to state $\tilde{\mathbf{s}}$ during the graph search
$L^{d,r}$	State ($\times$ time) lattice
m	Motion primitive
M	Size of time slice
N_{expl}	Number of exploration samples per bunch
N_{total}	Total number of samples per bunch
$\mathbf{p}$	Vector between critical points
p_{coll}	Overall probability for collision
$\bar{p}_{\text{coll}}(\tilde{\mathbf{s}})$	Collision probability for trajectory to state $\tilde{\mathbf{s}}$
p_{dyn}	Probability for collision due to dynamic obstacle
p_{stat}	Probability for collision due to static obstacle
p_{terrain}	Probability for collision due to terrain
$\mathbf{q}$	Configuration
$\mathbf{q}_{\text{g}}$	Goal configuration
$\mathbf{q}_{\text{s}}$	Start configuration
r	Resolution level
$\mathbf{r}$	Transformed vector between critical points
R_{g}	Goal radius

$\mathbf{R}$	Robot mask
$\mathbf{s}$	(Time-augmented) state
$\tilde{\mathbf{s}}_0$	Discrete start state of motion primitive
$\tilde{\mathbf{s}}_\mathrm{e}$	Discrete end state of motion primitive
$\mathbf{s}_g$	Goal state
$\mathbf{s}_\mathrm{s}$	Start state
$\tilde{\mathbf{s}}^{d,r}$	Discrete state, lattice point of $L^{d,r}$
$\tilde{\mathbf{s}}_\mathrm{s}^{d,r}$	Discrete start state
S_i^r	Set of discrete values for state variable s_i with resolution r
t	Time
t_0	Initial time
t_f	Final time
$t(m)$	Duration of a motion primitive's trajectory
$\bar{t}(\tilde{\mathbf{s}})$	Duration of trajectory to state $\tilde{\mathbf{s}}$ during the graph search
T^r	Set of discrete time values with resolution r
$\mathbf{u}$	Input, control
v	Velocity
V^r	Set of discrete velocity values with resolution r
x	Position (x-coordinate)
X^r	Set of discrete x-position values with resolution r
y	Position (y-coordinate)
Y^r	Set of discrete y-position values with resolution r

Greek letters

α	Weighting factor
β	Steering angle (input)
γ	Decay rate for collision probability
δ_x^r	Increment for variable x at resolution level r
Δ_t	Temporal quantization for collision checking
Δ_{xy}	Resolution of robot mask and time slice
Δt_m	Duration of a motion primitive

ϵ	Heuristic inflation factor
ϵ'	Suboptimality bound
ϵ_{d}	Relative cost increase for decomposition
ζ	Enlarged footprint radius
η_{b}	Penalty for driving backwards
η_{r}	Weighting factor for collision risk
η_{t}	Weighting factor for trajectory duration
θ	Heading
Θ^r	Set of discrete heading values with resolution r
κ	Kinematical constant
$\boldsymbol{\lambda}^r(\mathbf{s})$	Nearest lattice point for state $\mathbf{s}$
Λ	Lattice
μ	Metric
ν	Upper bound on overestimation of heuristic distance
ξ	State to configuration mapping
π	Projection
ρ	Minimum distance to next obstacle
Σ	Covariance matrix of dynamic obstacle
τ	Path parameter
τ_d	Temporal threshold for dimensionality level d
ϕ	Path
ϕ^*	Optimal path
ϕ_{steer}	Steering policy
ϕ^*_{steer}	Optimal steering policy
Φ	State evolution
ψ	Multi-resolution trajectory with hybrid dimensionality
Ω	Time slice

Miscellaneous

$\mathcal{A}$	Robot
$\mathcal{B}^{d,r}$	Motion primitive bunch
$\mathcal{C}$	Configuration space
$\mathcal{C}_{\text{free}}$	Free space
$\mathcal{C}_{\text{g}}$	Set of goal configurations
$\mathfrak{d}$	General heuristic distance
$\mathfrak{d}_{\text{Eucl}}$	Euclidean heuristic distance
$\mathfrak{d}_{\text{Lik}}$	Likhachev's heuristic distance
$\mathfrak{d}_{\text{obst}}$	Improved obstacle-aware heuristic distance
$\bar{\mathfrak{d}}_{\text{obst}}$	Threshold for computation of obstacle-aware heuristic distance
$\mathcal{M}^{d,r}$	Motion primitive set
$\mathbb{N}_0$	Set of natural numbers (including 0)
$\mathbb{N}^+$	Set of natural numbers (excluding 0)
$\mathcal{O}$	Obstacle set
$\mathbb{R}$	Set of real numbers
$\mathbb{S}^1$	Unit circle
$\mathcal{S}$	State space
$\mathcal{S}_{\text{g}}$	Set of states for goal g
$\tilde{\mathcal{S}}_{\text{g}}$	Set of discrete states for goal g
$\mathfrak{t}$	Heuristic remaining duration
$\mathcal{T}$	Set of continuous points in time
$\mathscr{T}$	Continuous-state trajectory
$\mathcal{U}$	Input space
$\mathcal{W}$	World
$\mathfrak{w}_i$	Minimum distance between goal g_{i-1} and g_i
$\mathbb{Z}$	Set of integers

Chapter 1

Introduction

Since their successful application for exploration tasks in disaster areas [Mur04; Nag13], mobile robots have established themselves as a valuable assistance for rescue teams. In the majority of cases, those robots are teleoperated and require a high level of attention from well-trained operators. Capabilities for (semi-) autonomous task execution for search and rescue robots can support the operators and are thus subject of active research. The R&D project SENEKA[1], which was funded under the Fraunhofer research program *Markets Beyond Tomorrow*, aimed at the development of concepts for networked heterogeneous robot teams [Kun14]. Mobile robot motion planning in an unknown, changing environment was a major topic within the SENEKA project. The majority of the research and development in this thesis was conducted in the context of this project.

At least since the Defense Advanced Research Projects Agency (DARPA) has provided substantial funding for participants of their competitions for self-driving cars (DARPA Grand Challenge in 2005 [Iag06] and DARPA Ur-

1. SENEKA: Sensornetzwerk mit mobilen Robotern für das Katastrophenmanagement (Engl. sensor network with mobile robots for disaster management).

ban Challenge in 2007 [Bue08]), mobile robot motion planning is an active field of research. Consequently, the main focus is on research for robotic vehicles driving on roads. This direction of research is also strongly promoted by vehicle manufactures (for an example see Daimler's technology demonstrator Bertha [Zie14]). Accordingly, specific aspects like cooperative driving [Fre11] or energy-efficient driving [Gua15] are common topics of research. Driving on roads has the big advantage that the basic path is known in advance because it is predetermined by the course of the road. This is why motion planning for mobile robots in urban areas generally relies on precise maps of the environment, which need to be built prior to driving [Zie14].

The scenarios considered in this thesis are entirely different: It is assumed that no map is available or that existing maps are no longer valid due to—possibly catastrophic—changes in the environment. Therefore, it does not suffice to merely plan a velocity profile and which lane of the road to use. Instead, the robot needs to find its way to the goal in an unstructured environment. This corresponds to the classical path planning problem; however, in addition, it is also assumed that the robot acts in a dynamic environment, which is populated by other vehicles or people. Conventional approaches (see Section 2.3) use hierarchical, decoupled methods for motion planning in unstructured dynamic environments: in the first phase, only a path to the goal is computed; in a secondary phase, a trajectory or velocity profile that tries to track the previously planned path is computed. If the planned motion leads to a collision with a dynamic obstacle, the trajectory is locally modified in order to let the robot slow down and let the obstacle pass or to perform a local avoidance maneuver. The local modification is generally performed without considering global optimality, which might lead to suboptimal results of the overall motion.

This thesis presents an approach that integrates global path planning and local obstacle avoidance in a unified, consistent motion planning algorithm without the separation of path planning and trajectory generation. Considering dynamic obstacles requires planning in a high-dimensional state × time space, which essentially amounts to solving an optimal control problem (OCP). Besides the optimization of design parameters of robotic mechanisms (cf. [Mil10]), the calculation of optimal motion plans is an often-encountered optimization problem in robotics. The efficient planning of optimal trajectories for nonholonomic robots, i.e., systems with differential constraints, is still an interesting field of research in robotics with many open questions—especially for planning in dynamic environments (see Section 2.8).

Calculating a high-dimensional optimal trajectory is computationally expensive, even more so, when planning over a long distance. Even with today's computing power, a comprehensive solution of the global optimization problem in unstructured environments is impracticable for real-time operation. This thesis presents a method that is based on the observation that it is not necessary to plan a full-dimensional high-fidelity trajectory up to the goal because the relevance of the individual state variables does not remain constant for the whole planning horizon. Instead, the relevance decreases with increasing distance and time. This applies to the relevance of the state variables themselves as well as to the resolution of each state variable (see Figure 1.1).

The algorithm proposed in this thesis performs full-dimensional high-fidelity time-parametrized motion planning in the vicinity of the robot (both spatially and temporally). A subsequent segment drops the time-parametrization and thus the consideration of dynamic obstacles; however, it still uses the full system state for planning, which allows the consideration of the system's dynamics. The final segment performs only path planning without considering derivatives of the configuration variables. This still allows to take kinematical constraints (like nonholonomicity) into account and ensures that the planned path can be tracked by the robot.

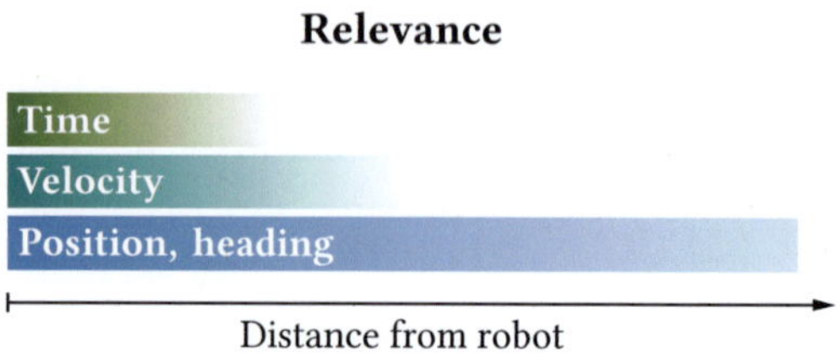

Figure 1.1: Relevance of planning fidelity. The state variables corresponding to the robot's position and heading are relevant from the start to the goal in order to guarantee a kinematically feasible solution; however, their resolution might be chosen more coarse with increasing distance from the robot. Considering velocity and time is especially important in the vicinity of the robot for planning among dynamic obstacles.

This thesis aims at the development of a holistic extensible approach that integrates global path planning and local time-parametrized motion planning. It supports additional features like the incorporation of terrain characteristics and combined planning along multiple waypoints in a consistent way without the need for the differentiation of special cases. The method is derived using the example of mobile robot motion planning; however, the algorithm is stated in a generic way and is applicable to a wide range of robotic systems.

1.1 Problem Statement

In order to lay out the groundwork for the following chapters, this section formally states the robot motion planning problem. The notation closely follows common conventions widely used throughout the literature, e.g., [LaV06]. This section concentrates on the definitions most specific to this thesis. Particular attention is paid to the differentiation between path planning, kinodynamic motion planning, and time-parametrized motion planning as the authors cited use slightly different terminology.

Each robot operates in a given environment, which is also commonly referred to as its *world* $\mathcal{W}$. Due to a broad variety of mapping algorithms, there exist a lot of different representations for this world (such as 2D occupancy grids [Thr96; Thr06a; Emt12], 3D Normal Distribution Transform maps [Mag07], cost maps based on drivability [Neu09], etc.). In its most general form, the world is composed of a region $\mathcal{O} \subset \mathcal{W}$ occupied by obstacles and the remaining free space $\mathcal{W} \setminus \mathcal{O}$.

The configuration $\mathbf{q}$ of the robot completely determines the position of each point of the robot and the set of all possible configurations is called the configuration space $\mathcal{C}$. The robot is denoted with $\mathcal{A}(\mathbf{q})$, which maps the robot's configuration $\mathbf{q} \in \mathcal{C}$ to the part of the world $\mathcal{W}$ occupied by the robot for this particular configuration. A configuration may, for example, consist of the joint angles for a robotic arm or of the position and heading for a mobile robotic platform. The robot is said to be in a collision whenever

$$\mathcal{A}(\mathbf{q}) \cap \mathcal{O} \neq \varnothing\,. \tag{1.1}$$

1.1.1 Path Planning

The basic path planning problem consists of finding a continuous transition from a given start configuration $\mathbf{q}_s$ to a goal configuration $\mathbf{q}_g$ or even an entire set of goal configurations $\mathcal{C}_g$. No part of the planned transition must be in a collision with any obstacle, i.e., all admissible configurations must be part of the free space[2]

$$\mathcal{C}_{\text{free}} := \{\, \mathbf{q} \mid \mathcal{A}(\mathbf{q}) \cap \mathcal{O} = \varnothing \,\}\,. \tag{1.2}$$

The temporal evolution of the robotic system (including its dynamics) is not considered during mere path planning. Thus, the basic problem

2. Although, strictly speaking, $\mathcal{C}_{\text{free}}$ is the free *configuration* space, it is commonly referred to just as *free space*.

of planning a path ϕ can be formally stated as follows [LaV11a]: Find a continuous mapping

$$\begin{aligned} \phi : [0, 1] &\rightarrow \mathcal{C}_{\text{free}} \\ \tau &\mapsto \mathbf{q} \end{aligned} \tag{1.3}$$

with $\phi(0) = \mathbf{q}_s$ and $\phi(1) \in \mathcal{C}_g$. For the most basic form of path planning, *any* path satisfying this condition constitutes a valid solution; however, given a specific application, it might be beneficial to search for an *optimal* solution ϕ^* by minimizing an appropriate cost functional J:

$$\phi^* = \underset{\phi(\tau)}{\arg\min}\ J(\phi(\tau))\,. \tag{1.4}$$

1.1.2 Kinodynamic Motion Planning

For systems with high dynamics or differential constraints, planning a path in configuration space only is often not sufficient. Although the terms *path planning* and *motion planning* are often used interchangeably by many authors, in the context of this thesis, the term *motion planning* is used to express the explicit consideration of a dynamic robot motion model during planning. This is also referred to as *kinodynamic planning* [Don93].

To include the additional constraints, a transition from the configuration space $\mathcal{C}$ to the state space $\mathcal{S}$ has to be made. A state vector $\mathbf{s} \in \mathcal{S}$ can be regarded as the generalized extension of the configuration vector $\mathbf{q}$ and its derivatives,

$$\mathbf{s} = \begin{bmatrix} \mathbf{q} \\ \dot{\mathbf{q}} \\ \ddot{\mathbf{q}} \\ \vdots \end{bmatrix}, \tag{1.5}$$

with an accompanying mapping ξ that maps the state $\mathbf{s}$ to its associated configuration $\mathbf{q}$,

$$\begin{aligned} \xi : \mathcal{S} &\to \mathcal{C} \\ \mathbf{s} &\mapsto \mathbf{q}\,. \end{aligned} \tag{1.6}$$

With (1.5) the dynamics of the robot can be modeled as a generic nonlinear control system

$$\dot{\mathbf{s}}(t) = \mathbf{f}(\mathbf{s}(t), \mathbf{u}(t)) \tag{1.7}$$

where $\mathbf{u} \in \mathcal{U}$ is the vector of input (i.e., control) variables. This intuitively leads to the definition of the general motion planning problem: Find a steering policy ϕ_{steer} that moves the dynamic system (1.7) from its initial state $\mathbf{s}_{\text{s}}$ to a goal state $\mathbf{s}_g \in \mathcal{S}_g$ while avoiding any collision. This can be formally stated—following [LaV06]—as

$$\begin{aligned} \phi_{\text{steer}} : [0, t_{\text{f}}] &\to \mathcal{U} \\ t &\mapsto \mathbf{u} \end{aligned} \tag{1.8}$$

with its accompanying state trajectory

$$\mathbf{s}(t) = \mathbf{s}(0) + \int_0^t \mathbf{f}(\mathbf{s}(\tau), \mathbf{u}(\tau))\,\mathrm{d}\tau \tag{1.9}$$

and subject to

$$\begin{aligned} \mathbf{s}(0) &= \mathbf{s}_{\text{s}}\,, \\ \mathbf{s}(t_{\text{f}}) &\in \mathcal{S}_{\text{g}}\,, \end{aligned} \tag{1.10}$$

and

$$\forall t \le t_{\text{f}} \quad \xi(\mathbf{s}(t)) \in \mathcal{C}_{\text{free}}\,. \tag{1.11}$$

Similar to the path planning problem, basic motion planning does not impose any constraints on optimality, and especially the final time t_f may be chosen arbitrarily. However, in many applications an optimal solution with respect to a given optimality criterion is often desired. For this purpose, the motion planning problem can be converted to the optimization problem

$$\phi^*_\mathrm{steer} = \underset{\mathbf{u}(t),\,t_\mathrm{f}}{\arg\min}\; J(\mathbf{s}(t), \mathbf{u}(t), t_\mathrm{f}) \tag{1.12}$$

constrained by (1.10) and (1.11). This formulation amounts to the classical definition of an OCP with the additional free space constraint (1.11). Thus, the general motion planning problem can be regarded as a special case of OCPs.

1.1.3 Time-Parametrized Motion Planning

So far, the definitions of the path and motion planning problem both have assumed a static environment for the robot. However, especially beyond traditional industrial applications, this assumption is unrealistic. For example, sensors may gather additional information about the environment while the robot is moving, or the environment itself may be dynamic because of moving obstacles. In order to distinguish between motion planning in a static environment and planning in a dynamic environment, the latter is referred to as *time-parametrized motion planning* in this thesis.

The formal definition of the time-parametrized motion planning problem is identical to the definition of the motion planning problem in Section 1.1.2 with one important exception: the time-invariant free space $\mathcal{C}_\mathrm{free}$ becomes the time-variant free space $\mathcal{C}_\mathrm{free}(t)$, and thus the problem of time-parametrized motion planning can be stated as

$$\begin{aligned} \phi^*_\mathrm{steer} &= \underset{\mathbf{u}(t),\,t_\mathrm{f}}{\arg\min}\; J(\mathbf{s}(t), \mathbf{u}(t), t_\mathrm{f}) \\ \text{s.t.} \quad & \forall t \leq t_\mathrm{f} \quad \xi(\mathbf{s}(t)) \in \mathcal{C}_\mathrm{free}(t)\,. \end{aligned} \tag{1.13}$$

1.2 Scope and Objectives

This thesis is concerned with time-parametrized motion planning in unstructured dynamic environments. As already mentioned in the beginning of the introduction, this differs from planning motions in on-road scenarios because unstructured environments lack the a priori information of a basic path to the goal and require to take the terrain characteristics into account. This thesis aims at the development of a unified motion planning concept that integrates global path planning and local trajectory planning in a consistent way. The idea is to guarantee robust avoidance of collisions with dynamic obstacles while still having the optimality of the overall global plan to a possibly distant goal in mind. The motion planning algorithm should be able to generate resolution-optimal plans with respect to a user-specified cost function. Thus, the proposed method will provide an approximate solution for the time-parametrized motion planning problem (1.13), for which only local continuous solutions can be computed; all methods for finding a globally optimal solution need to employ some form of sampling [How07]. Furthermore, the proposed algorithm is also resolution-complete, which means that it will always return a solution if such a solution exists (assuming a given resolution); or on the other hand, if no solution exists, the algorithm will report this fact in finite time.

Besides performing genuine time-parametrized motion planning in the vicinity of the robot for the consideration of dynamic obstacles, the algorithm is able to cope with a wide range of robot kinematics (e.g., kinematics with nonholonomic constraints). In addition, the algorithm intends to be as generic as possible and allows the straightforward integration of further capabilities. This is shown using the example of planning along multiple waypoints or taking terrain characteristics into account. In order to be suitable for real-world operation, i.e., to be able to react to sudden changes in the environment, the algorithm should be capable of computing solutions with a rate of at least 10 Hz.

This thesis focuses on the solution of the motion planning problem. Of course, autonomous driving requires the availability of many more components like mapping of the environment (e.g., [Gri05; Mon07]), localization (e.g., [Mou06; Agh13]), and trajectory tracking and execution (e.g., [Lam08; How10]). These are all wide-ranging research fields of their own and are not part of this thesis. Nonetheless, some of them had to be touched on to allow a meaningful evaluation of the proposed planning concept.

1.3 Scientific Contributions

The key contribution of this thesis is a novel holistic concept that integrates local maneuver planning in the presence of dynamic obstacles and global path planning in a unified consistent approach. The proposed method is not a composition of existing algorithms with individual capabilities; instead, it is one single algorithm that is designed to make the most of the available information. The approach converts the continuous OCP to a search for the shortest path in a graph in order to be able to draw on the large body of established graph search algorithms. Thus, the main contribution of this thesis is a method to construct this search graph in the form of a hybrid-dimensional multi-resolution state × time × goal lattice. The individual contributions are as follows.

- A method for systematic **successive dimensionality reduction** has been developed to reduce the computational complexity [Pet14]. To consider dynamic obstacles, planning starts in a full-dimensional state × time lattice. At a certain threshold, planning continues in a high-dimensional state lattice that is still capable of correctly representing the robot's dynamics. The last part of the planning produces only paths, which do, however, still respect the kinematical constraints of the robot.

- In order to reduce the computation time even further, **multi-resolution planning** has been integrated into the holistic concept [Pet13b]. For this purpose, a new method that uses the map of the environment to derive the planning resolution has been developed.

- The edges of the search graph are made up of motion primitives. An offline method for the **sampling of motion primitives from arbitrary system models** has been developed [Pet13b; Pet14]. The method is based on the minimization of the quantization error between lattice points and the end points of the motion primitives. In addition, motion primitive sets for state lattices with lower dimensionality are derived from the full-dimensional motion primitive set.

- The planning algorithm has been extended with **combined planning along multiple waypoints** [Pet13c]. This is also directly integrated into the structure of the search graph. The combined planning enables the robot to arrive at a waypoint in an optimal pose to reach its next destination.

- For the efficient application of existing graph search algorithms, **heuristics suitable for multi-waypoint planning** have been developed [Pet13c]. In addition, a Dijkstra-based 2D heuristic has been improved in order to be more informative than the existing method is.

- For time-parametrized planning in the vicinity of the robot, a method for **probabilistic modeling of dynamic obstacles** has been developed. For this purpose, an existing algorithm for circular obstacles has been extended with caching capabilities [Pet13c], and a novel algorithm, based on time slices, for obstacles with arbitrary shape has been devised.

The proposed algorithm has been thoroughly evaluated and a comprehensive analysis and discussion is provided in this thesis. In order not to be restricted to simulation results, the algorithm has also been implemented on a robotic platform and was tested in real-world operation [Kun14].

1.4 Thesis Structure

So far, **Chapter 1** has stated the basic problem of robot motion planning in a dynamic environment and the objectives of this thesis together with a brief outline of the proposed planning approach. **Chapter 2** summarizes the results of previous research in order to put the proposed planning concept into relation with the state of the art in motion planning. Special attention is given to the open issues in the field. **Chapter 3** presents the novel holistic planning concept. It is explained how a state × time lattice can be defined for a given system model and how the corresponding motion primitives can be computed. The chapter deductively describes how the initial state × time lattice is used to derive further state lattices with decreased dimensionality and variable resolution (once again with the corresponding motion primitives). Additionally, it is shown how the state × time lattice can be extended to a state × time × goal lattice to allow for combined planning over multiple waypoints. Although the application of the proposed concept is not restricted to *mobile* robots, the concept is illustrated using the example of a four-wheeled robot, which is also revisited later on for the evaluation of the results. **Chapter 4** is a brief interlude that describes how the environment and especially dynamic obstacles are modeled to allow for realistic testing of the proposed planning algorithm both in simulation and real-world operation. **Chapter 5** shows how the state lattices and motion primitives from Chapter 3 can be used to construct a search graph which allows the conversion of the general motion planning problem to the problem for finding the shortest path in a graph, for which well-established algorithms exist. One of these algorithms, namely Anytime Repairing A* (ARA*), which is able to iteratively improve the solution, is explained in detail including necessary modifications and extensions. **Chapter 6** presents the results which have been obtained both from simulations based on real recorded environment data and from application of the algorithm on a physical robotic platform. This allows for a comprehensive evaluation of the proposed planning concept and comparison with other algorithms. Finally, **Chapter 7** summarizes the contributions of this thesis and discusses potential future work.

Chapter 2

State of the Art

Since the beginning of industrial robotics in the late 1970s, planning algorithms for robotics have been subject of extensive research. Early work was strongly limited to mere path planning problems within the industrial context. It was mostly of a theoretical nature without practical applicability to challenging tasks. The limited processing power rendered the consideration of realistic robot and environment models impossible.

As computer technology evolved, research in robot planning algorithms gathered pace, which led to a huge amount of publications and some very comprehensive books that attempt to organize the topic and compile the various methods [Lat91; Lau98; Cho05; LaV06].

Today, the problem of robot motion planning for industrial standard applications can be considered as solved. However, in mobile robotics—especially for autonomous vehicles operating in unstructured and dynamic outdoor environments—many open questions still exist. This is the reason why ongoing research is conducted in this field. The substantial funding (e.g., Defense Advanced Research Projects Agency (DARPA) Grand Challenge [Iag06] and Urban Challenge [Bue08]) shows the particular need for the development of improved techniques. Automotive companies see autonomous driving as a future key technology and have established corresponding research departments.

This chapter gives an overview of the current state of the art in mobile robot motion planning. First, robot path and motion planning in general is briefly covered in order to lay the foundations for the more advanced topics. Following this, the results of the latest research which focus on mobile robot motion planning in unstructured dynamic environments are presented. Where applicable, the particular advantages and limitations of the presented methods are discussed.

2.1 Early Planning Approaches

Robot path planning by means of visibility graphs [Loz79] was one of the first methods for finding collision-free paths among polyhedral obstacles. The method computes shortest paths by design as the resulting path consists of straight line segments connecting the obstacle corners. However, for safety reasons, it might not be desirable for the final path to touch the obstacles. Therefore, algorithms based on Voronoi graphs have been proposed [ÓDú85]. They maximize the clearance between robot and obstacles by tying the planned path to the Voronoi diagram of the planning space.

Cell decomposition methods [Cha87] convert the planning space to a very reduced representation while maintaining topological equivalence. The resulting graph can then be searched using standard shortest path algorithms.

Artificial potential fields [Kha86] are a completely different approach that models the robot as a particle moving in a potential field. Obstacles exert a repelling force on the particle while the goal is modeled by an attracting force. The basic version of this algorithm is prone to getting stuck in local minima and thus only works if all obstacles are convex.

All these methods are unable to incorporate a model of the robot's kinematics or dynamics during the planning. They only consider the simple path planning problem from Section 1.1.1 in unrealistic polygonal environments.

2.2 Path vs. Motion Planning

Motion planning as defined in Section 1.1.2 is computationally expensive. Therefore, a popular approach is to plan a path first and then compute a suitable velocity profile [Kan86]. Although this method of decoupled position and velocity planning clearly is not able to guarantee optimality or completeness, it works reasonably well in practical applications and is still frequently employed, e.g., during both the DARPA Grand Challenge by the CMU team [Urm06] and the DARPA Urban Challenge by team AnnieWAY [Kam08].

Donald et al. were one of the first to systematically examine the differences between path and motion planning [Don93]. The authors introduced the term *kinodynamic motion planning* for planning problems that have to deal with both kinematic and dynamic constraints. According to their definition, kinematic constraints limit the configuration of the robot (e.g., obstacles and joint limits); dynamic constraints limit the respective time-derivatives (e.g., bounded velocity or acceleration). Donald et al. also introduced a third category of constraints, called *strictly kinodynamic* constraints, to describe dynamic constraints that depend on kinematic properties (e.g., speed-dependent obstacle-avoidance margin). Essentially, the proposed method performs a breadth-first search in a regular grid embedded in the state space.

For motion planning in a dynamic environment, most often local planning strategies (see Section 2.3) are employed. *Velocity obstacles* [Fio98] are a widely used concept, that models the potential collision region of a robot and obstacle on the basis of their relative velocity vector. While the original version was limited to linear obstacle motions only, it was later extended to cope with nonlinear obstacle trajectories [Lar02; Shi07]. However, all variants are limited to both circular robots and obstacles, so other shapes have to be approximated by multiple circles.

In [Lin05; Lin09] the authors propose a method for smooth feedback planning by exploiting a combination of cell decomposition and artificial potential fields, called *global vector field*. The basic version, being limited to holonomic point robots, was extended to car-like robots in [Lin07]. Both variants operate in a polygonal environment, for which efficient cell decomposition methods exist, but unfortunately are difficult to apply to unstructured, possibly dynamic environments.

As stated in Section 1.1.2, motion planning can also be thought of as a special case of optimal control. For planning the motion of a Mars rover in rough terrain, [How07] converts the motion planning problem to a nonlinear programming (NLP) problem. It comprises a complex dynamical model for a wheeled mobile robot. The solver relies on good initial values, which are obtained from a look-up table[1].

A partially closed-loop receding horizon control is proposed in [DuT10] for motion planning in dynamic, cluttered, uncertain environments. However, it is not clear whether the proposed method is real-time capable. Generally speaking, motion planning and the subsequent robot control can be thought of as nonlinear model predictive control (NMPC) [How14]. However, the cycle time in robotic applications is much shorter compared to the well-established NMPC application field of process control [Pet10].

In order to cope with the demanding real-time requirements of robotic motion planning, it might be a viable strategy to compute a suboptimal plan first and then further optimize the trajectory in an additional processing step [Bet98].

1. A *look-up table* is a predefined/precomputed data set "library".

2.3 Global vs. Local Planning

Another widely used approach for keeping the computational burden low is the separation of the planning problem into global path planning and local motion planning / obstacle avoidance. The global path usually does not consider kinematic or dynamic constraints and is planned with a moderately low rate. A subsequent local motion planning algorithm tries to track the global path and is executed at a higher rate in order to react to changes in the environment. This separated approach sacrifices global optimality in favor of local responsiveness.

The curvature-velocity method [Sim96] models the planning problem as a constraint optimization in the velocity space and is closely related to the local Dynamic Window Approach (DWA) [Fox97]. In order to guide the local planning in the direction of the global optimum, the DWA was augmented with the global navigation function NF1 [Bar91b] in [Bro99] for holonomic robots. [Phi03] combined the DWA with Elastic Bands [Qui93] for a differential-drive tour guide robot. On the basis of DWA, [Phi07] proposed the E* algorithm for car-like robots. It exploits a new navigation function that is aware of environment characteristics like traversability risk.

In [Shi91] the authors propose an algorithm that consists of two stages: First, multiple global path candidates are computed; secondly, the most promising global path is selected and further optimized locally. A similar approach is presented in [Reu98], where trajectory generation is performed by solving an optimal control problem in order to continuously track a previously computed global path.

Sometimes, it might even be possible to spare the global planning. For example, the Stanford Racing Team, who won the DARPA Grand Challenge, relied on a local planner only, because an exact geometrical description of the road-network was available [Thr06b]. Their local planner (as well as many others, e.g., [Ger08]) is based on the simulation of a set of many local trajectory candidates, among which the best is chosen on the basis of an appropriate optimality criterion. This approach is also known as *trajectory rollout*.

The hierarchical motion planning proposed in [How08] led to the algorithm used for *Boss* in the DARPA Urban Challenge [Fer08a]. This two-layered approach consists of a high-speed trajectory generator (described in [How07]) for on-road driving and a state lattice planner with subsequent trajectory generator for driving in unstructured areas like parking lots. Furthermore, hierarchical planning concepts have been successfully applied in the context of mobile manipulation [Kne10].

For dynamic street scenarios, local trajectory generation has been developed in [Wer10; Wer11]. The algorithm simultaneously maximizes driving comfort (minimum jerk) and tracking performance using optimal control techniques. The authors of [Zie15] underline that continuous local methods may generally get stuck in local minima which is why they propose the transformation of their variational method to a method based on Hidden Markov Models for global trajectory planning. Moreover, Partially Observable Markov Decision Processes (POMDPs) have been used to devise planning algorithms that explicitly take both perception and action uncertainty into account; however, in real-world applications only approximate methods can be used for planning with POMDPs due to the high complexity of the planning problem [Thr06a].

2.4 Nonholonomic Path Planning

Planning under differential constraints is still a major field of research. If the number of action variables is less than the dimensionality of $\mathcal{C}$, a system is called to be underactuated [LaV11b]. Most mobile robots are nonholonomic systems due to the rolling wheels that constrain the possible velocities of the robot with regard to its instantaneous configuration.

One of the first papers to address nonholonomic planning problems was [Lau87], which demonstrated that fundamentally new methods needed to be introduced to tackle nonholonomic planning problems. First approaches modified existing methods in order to cope with the new requirement. For example, [Jia92] extended the visibility graph method to a kinematically feasible visibility graph.

Most nonholonomic planning algorithms can be split into two separate steps. First, a relatively simple and often holonomic solution is planned. Secondly, this initial solution is modified in order to satisfy the nonholonomic constraints.

In order to support the modification phase, [Sek99] proposed an extension to artificial potential fields by incorporating a nonholonomic metric. This results in an initial solution that is easier to modify with regard to the nonholonomic constraints. This modification might be performed by recursively subdividing the initial path and trying to connect the end points by Reeds-Shepp shortest paths [Ree90] and a subsequent path optimization [Lau94; Lam01]. Some authors have proposed methods to progressively introduce kinematic constraints until the path correctly represents the nonholonomic robot motion [Fer98]. An algorithm for tractor-trailer systems with multiple nonholonomic constraints was presented in [Sek98]. It consists of several planning iterations: First, a solution considering only one nonholonomic constraint is planned. Then, the path is iteratively "repaired" by adding one additional constraint at a time. For this purpose, the authors propose two methods: pick and link (similar to [Lau94]) and a probabilistic planning in a tube along the path.

Obviously, a two-staged approach with initial unconstrained global path planning and subsequent local modification and optimization cannot provide any guarantees on global optimality or completeness. Furthermore, the above-mentioned algorithms deal with path planning only. They do not address the general motion planning problem with a complete incorporation of the system's dynamics.

2.5 Sampling-Based Planning

Sampling-based planning algorithms try to reduce the complexity of a planning problem by probabilistically sampling configurations from a (possibly high-dimensional) configuration space instead of enumerating every possible configuration. The sampling process attempts to rapidly explore the entire search space while capturing its topology as accurate as possible. The improvement in computation time comes at the expense of solution quality. The result is non-deterministic and generally not optimal. The algorithms are only probabilistically complete, i.e., they can only guarantee to find an existing solution if the sample count tends to infinity.

Nevertheless, sampling-based planning algorithms have proved their worth and are backed by a large community. There exist several open-source projects that provide free implementations of well-known and established algorithms, the most popular being the Open Motion Planning Library (OMPL) [Şuc12b], which is extensively used by *MoveIt!* [Chi12], the main planning framework of the Robot Operating System (ROS).

The two major groups of sampling-based planning algorithms, namely Probabilistic Roadmaps and Rapidly-exploring Random Trees are discussed in the following sections.

2.5.1 Probabilistic Roadmaps

Planning with Probabilistic Roadmaps (PRMs) consists of two phases: First, a configuration space roadmap is constructed by randomly sampling configurations and trying to connect these configurations using a local planner. Secondly, the graph constructed in phase one is queried for the shortest path between a particular start and goal configuration [Kav96]. Originally developed for holonomic planning problems, the method was extended to car-like robots in [Šve97].

PRM methods have difficulties with narrow passages. These difficulties may be overcome by a temporary dilation of passages to capture the connectivity of the configuration space [Hsu98].

In order to reduce the roadmap construction time, Lazy PRM algorithms have been proposed [Boh00]. They defer collision checking to the query phase, whereas the original PRM algorithm performs the collision checking during roadmap building. Therefore, Lazy PRM is especially suited for single query problems. To further improve the planning performance, a bi-directional Lazy PRM method has been proposed [San01]. It simultaneously uses two search trees rooted at the start and goal configuration.

Finding a good sampling strategy is the major challenge for PRM-based algorithms. Quasi-random PRM (Q-PRM) extends Lazy PRM with a low-discrepancy and low-dispersion sampling strategy [Bra01]. This sampling of configurations in a regular pattern leads to a faster configuration space coverage and avoids the clustering of the samples. Because of the improved sampling strategy, the algorithm becomes resolution-complete, i.e., if a solution exists for the specified resolution, the algorithm is guaranteed to find it. An iterative refinement to this sampling strategy was introduced in [Lin03] by exploiting an incremental low-discrepancy lattice. Although it was initially thought that the strength of PRMs lies in the probabilistic sampling approach, research showed that random sampling is not advantageous over deterministic sampling [LaV04].

Originally developed for planning in configuration space only, PRMs have been extended for planning in state × time space [Hsu02]. For this purpose, a tree of (state, time)-tuples (called milestones) is grown by randomly applying controls to randomly selected milestones of the tree.

A method for efficient roadmap updating was proposed in [Jai04] for dynamic environments which are mainly static but contain a relatively small changing region. The algorithm presented in [Şuc12a] improves the search space exploration strategy by employing an underlying grid to keep track of the search space coverage.

All PRM methods depend on the availability of an efficient local planner in order to connect two nodes of the roadmap. For path planning only, this local "connect" step can be performed by using geometrical primitives like Dubins or Reeds-Shepp paths [Dub57; Ree90]. A more advanced method adds continuous curvature to the local paths by composing them from straight line segments, circular arcs and clothoids [Fra04a]. For motion planning incorporating the robot's dynamics, the "connect" step becomes much more involved as each connection requires the solution of a boundary value problem.

2.5.2 Rapidly-Exploring Random Trees

Rapidly-exploring Random Trees (RRTs) [LaV98] mitigate the connection problem of PRMs. The operating principle of an RRT rooted at a given start configuration is as follows: First, a configuration is sampled from the free space and the nearest neighbor in the RRT is selected. Secondly, a control is applied to this selected configuration to move the system toward the previously sampled configuration ("extend" step). Finally, if no collision occurred, the resulting new configuration is added to the RRT. These steps are repeated until the tree connects to the goal configuration. RRTs are well suited for single query planning in high-dimensional spaces because the sampling strategy biases the growing of the tree toward unexplored regions of the search space. Like most sampling-based planning algorithms, RRTs are probabilistically complete but not asymptotically optimal.

A thorough analysis of RRTs was performed in [LaV00], which also proposed several improvements: Better sampling strategies focused the search toward the goal, and the authors experimented with various sizes of the "extend" step (extreme case: directly connecting the tree with the random sample). The authors provided a lot of application results for holonomic, nonholonomic, and kinodynamic planning problems in static environments.

The RRT-connect algorithm [Kuf00] extends the basic RRT with bidirectional planning. The advantages—especially in the context of kinodynamic planning—are discussed in depth in [LaV01]. A resolution-complete version of RRT was presented in [Che02] and exploits the neighborhood structure of the state space.

Similar to PRMs, there has been a trend toward derandomizing RRTs: Multi-Sample RRT was proposed in [Lin04] and effectively captures the Voronoi structure of the search space. However, the algorithm has difficulties in narrow corridors and is susceptible to local minima.

The authors of [Fra02] applied analytically derived optimal control policies (e.g., obtained by minimizing a quadratic cost function) to the "extend" step, that move the system to equilibrium points (i.e., states with zero velocity). In the DARPA Urban Challenge, this algorithm constituted the basis for Team MIT's vehicle *Thalos*, which employed Closed-loop RRT, an algorithm that samples reference inputs for closed-loop stable vehicle dynamics [Kuw08; Kuw09].

Anytime RRT was proposed in [Fer06a]. It quickly generates an initial solution and then iteratively refines the solution quality as long as the allocated computation time allows. Another approach for reducing the computational effort was explored in [Mor04] by applying sampling-based planning to discretized search spaces. The authors of [Vah08] proposed a variant of RRT with adaptive dimensionality for humanoid robots with many degrees of freedom.

For planning in dynamic environments multipartite RRTs [Zuc07] and methods based on Gaussian processes [Ful08] have been suggested.

A rigorous analysis of sampling-based planning algorithms was conducted in [Kar11]. The authors proved that PRMs and RRTs, although probabilistically complete, are generally not asymptotically optimal. Hence, the authors present two algorithms (namely, RRT* and PRM*) that are asymptotically optimal. However, they do not consider differential constraints or dynamic environments. The algorithms were extended to kinodynamic planning in [Kar10] and nonholonomic planning in [Kar13] but only for static environments. The algorithms depend on the availability of an optimal solution to the short-time steering problem. Since the authors mention planning times within the range of several seconds, it is not clear whether the proposed algorithms are real-time capable for mobile robot applications.

2.6 Search-Based Planning

Search-based planning, also known as combinatorial planning [LaV11a], is an alternative approach to sampling-based planning. Search-based planning works by (implicitly) enumerating all possible robot motions in order to determine the optimal motion with respect to a particular optimality criterion. This enumeration might be done either for the inputs or directly for configurations or states. The necessary discretization of the search space might lead to missed solutions if the quantization is chosen too coarse. On the other hand, in contrast to sampling-based planning algorithms, search-based planning algorithms are normally guaranteed to be resolution-complete: If, given the discretization of the search space, a solution exists, the algorithm will find it. If no solution exists, the algorithm will report that fact in finite time. Because of their deterministic nature, search-based algorithms deliver reproducible results.

For mobile robots, Barraquand and Latombe proposed a search-based planning algorithm that discretizes control variables and can account for nonholonomic constraints [Bar89]. The algorithm performs a best-first search by applying one control value at a time and simulating the robot's motion for a certain period. In order to reduce the computational costs, the authors map the configuration space to a grid structure to keep track of already explored regions. The algorithm was extended in [Bar91a; Bar93] for faster planning and forms the basis for the Hybrid A* algorithm (see Section 2.6.1). Despite being restricted to path planning only, the algorithm is very popular and still constitutes the key idea of many current algorithms (e.g., the multi-resolution planner proposed in [Lin06] and the off-road-restrictions-aware planner in [Lin08]).

A search-based motion planning method, which relies on piecewise constant acceleration input (called *acceleration bang motion*), was presented in [Fra93] for planning in state × time space.

Since search-based planning algorithms suffer from the curse of dimensionality, most approaches try to keep the planning dimensionality low. For this purpose, a two-layered planning concept has been developed in [Che99]: It first plans for sub-goals in a dimensionally reduced subspace (e.g., only position and heading) and then uses these sub-goals as input for a second, subsequent full-dimensional planning step considering the physical model of the robot. The method is not optimal and with planning times ranging in the magnitude of hours, the algorithms application is limited to systems with slow dynamics (e.g., Mars rover).

The use of motion primitives for path planning was introduced in [Lac98]: The planning is split into two phases, offline motion primitive generation (integration of system model) and online search by combining the previously computed motion primitives. To reduce the computation time, the authors suggest the use of low-fidelity trajectories in distant regions.

In order to reduce the computational complexity even further, the concept of motion planning in state lattices has been developed [Piv09c], which is discussed in more detail in Section 2.6.2. Another approach to speed the planning up was proposed in [Phi11] and hinges on the idea of compressing the time dimension by using so-called *safe intervals*. The basic algorithm can only use travel time as a cost function, but it was extended to arbitrary cost functions in [Gon12].

A method called *path set relaxation* was proposed in [Krü10]. It improves the solution quality by simultaneously optimizing the motion primitives during the graph search.

Variants of the above-mentioned search-based motion planning algorithms are used in a variety of mobile robot applications, e.g., for autonomous mine mapping [Bak04] and in the DARPA Urban Challenge by Team AnnieWAY [Kam08]. The Search-Based Planning Library (SBPL) [Lik16] provides building blocks for many search-based algorithms but focuses on path planning problems only.

2.6.1 Hybrid A*

Building on the ideas of [Bar91a; Bar93], a very successful path planning algorithm called Hybrid A* was proposed in [Dol08] and further detailed in [Dol10]. The algorithm exhibits a hybrid nature between discrete graph search and continuous planning, hence the name. The path is composed of simple motion primitives (circular arcs) with quantized heading change. These motion primitives are concatenated during an A*-like graph search. In addition, a grid for the translational robot states is used to keep track of already explored regions: When a node is added to the graph, the continuous robot position associated with that node is stored in the corresponding cell of the underlying grid (see Figure 2.1). This continuous position constitutes the starting point for the subsequent graph expansion. The Hybrid A* algorithm allows for the generation of smooth paths yet benefiting from the fast exploration speed of grid-based search algorithms.

The Stanford Racing Team successfully used Hybrid A* for their robot *Junior* in the DARPA Urban Challenge [Mon08]. They extended the algo-

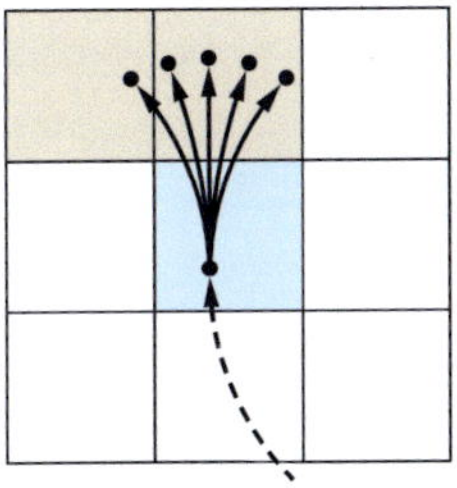

Figure 2.1: Operating principle of Hybrid A* (2D projection) [Pet12]: □ already explored cell, □ cells with newly added nodes.

rithm with multi-resolution capabilities, i.e., adaptive arc lengths, for faster planning and with a subsequent trajectory optimization step for smoother paths.

The Hybrid A* algorithm was further extended in [Pet12] particularly with regard to the special needs of off-road planning: Besides the efficient incorporation of terrain characteristics, simultaneous planning along multiple waypoints was introduced including the required advanced heuristic techniques.

2.6.2 State Lattices

Planning in state lattices is another successful method for mobile robot motion planning and was simultaneously developed by Pancanti et al. [Pan04] and Pivtoraiko and Kelly [Piv05b; Piv06]. While the former focused on the investigation of theoretical properties of the dynamical systems and provided only theoretical observations for polyhedral environments, the latter proved the applicability of state lattice planning to real-world motion planning problems. The technical details of the algorithm are comprehensively described in the corresponding technical report [Piv07], the journal article [Piv09c], as well as in Pivtoraiko's PhD thesis [Piv12].

Like many other motion planning algorithms, this algorithm is also composed of two separate steps: an offline motion primitive generation phase and an online graph search phase. The outline of the algorithm is as follows: First, the state space is discretized, which results in a lattice structure of the search space. Secondly, a set of kinematically feasible motion primitives is carefully crafted such that the start and end state of each motion primitive coincides with a lattice point of the state space. Finally, in the online planning phase, a graph is constructed by concatenating these motion primitives. The planning solution is obtained by finding the shortest path in this graph, where nodes represent states and edges represent controls, i.e., motion primitives.

The original algorithm considers only the general path planning problem: States consist of the robot's position, heading, and curvature in order to guarantee kinematically feasible paths. An extended algorithm, able to respect the robot's dynamics, was proposed in [Fer08a; Lik09] and field-tested on *Boss*, the robot that won the DARPA Urban Challenge [Urm08]. The planning is performed in a state space consisting of position, heading, and velocity, but no temporal dimension. Dynamic obstacles are thus represented as static risk zones appearing in a certain distance from the current robot position. Near the vehicle and near the goal a more densely sampled action space is employed resulting in basic multi-resolution capabilities of the algorithm. The graph search through the lattice is performed using the AD* algorithm [Lik05] together with a heuristic look-up table (HLUT) [Kne06]. The planned paths are finally tracked by an additional trajectory generation algorithm.

Several extensions have been developed to the original state lattice planner: A method to consider translational and rotational velocities was proposed in [Piv09a], which uses the ARA* algorithm [Lik03a] for efficient anytime planning. For on-road planning scenarios, the concept of deformable input-/state-lattice graphs has been introduced [Ruf10]. The algorithm is able to bend the lattice along an arbitrary continuously differentiable path. However, the solution quality may suffer from the non-uniformly sampled heading dimension and configuration-dependent steering angle quantization. Furthermore, the fact that the satisfaction of kinematic constraints has to be checked during the online planning may result in increased planning times.

Mobile robot motion planning using spatiotemporal lattices was first proposed in [Zie09] for on-road driving scenarios. The method utilized quintic splines for the geometric representation of path segments. Since planning in a spatiotemporal lattice results in searching a directed acyclic graph (DAG), an exhaustive search of the graph becomes possible. On the basis of this method and incorporating results from the work on time-parametrized planning in [Ruf09b], an algorithm was proposed in [McN11] with focus on short-horizon on-road driving scenarios exhibiting fast dynamics like leader following or swerving around dynamic obstacles.

2.6.3 Heuristic Graph Search Algorithms

Most search-based motion planning algorithms rely on the use of efficient graph search algorithms. This is a vast research field of its own and an exhaustive discussion is far beyond the scope of this thesis; however, in order to understand the motivation and implications of the graph search algorithm utilized in this thesis, a short overview will be given in the following.

For DAGs, which emerge in pure time-parametrized planning methods, the optimal solution can be computed with computation time linear in the number of vertices due to the topological ordering based on time [Zie09]. However, time-parametrized planning for the entire motion is only feasible for a relatively short time horizon as is the case, for example, in planning for on-road driving. In an unstructured environment, where no a priori information on the route to a distant goal is available, complete time-parametrized planning is computationally intractable. Unfortunately, as soon as the time dimension is removed from some nodes in the graph, cycles may occur. In this case, one must resort to generic algorithms that are capable to find shortest paths in arbitrary directed graphs.

The oldest and most known graph search algorithms are Dijkstra's algorithm for finding shortest paths [Dij59] and its heuristic derivative A* [Har68], which were invented already fifty years ago. Especially heuristic methods are very attractive for application to mobile robot motion planning because the structure of the state space with its embedded xy-plane allows for a simple and intuitive definition of a well-informed heuristic based on the Euclidean distance to the goal. The term *heuristic search* is somewhat misleading: heuristic graph search algorithms always find the optimal solution; the heuristics are only used to speed up the planning by guiding the search toward the goal. In fact, it can be even proved that A* is *optimally efficient* for any given heuristic function, which means that there is no other optimal algorithm that is guaranteed to expand fewer nodes than A* [Rus95].

After a long time of moderate activity, research in the field of heuristic graph search algorithms for robot motion planning applications has gained pace in the decade and brought a plethora of new algorithms. A comprehensive overview can be found in the survey "Heuristic search comes of age" [Stu12]. Besides static planning like A*, heuristic graph search algorithms for robot motion planning can be roughly categorized into incremental replanning algorithms, anytime algorithms, and anytime replanning algorithms [Fer05].

Incremental replanning algorithms

Incremental replanning algorithms, often just called replanning algorithms, are designed for repeatedly planning paths in a changing environment. After the initial planning phase, in each planning cycle, the algorithms try to repair the solution obtained during the previous planning cycle in order to make it consistent with the changed graph. If only a small fraction of the graph has changed during two planning cycles, this repairing may be considerably faster than planning from scratch. Popular members of this class of algorithms are the D* [Ste94] and D* Lite [Koe05] algorithms, where "lite" refers to the simpler implementation of the latter without sacrificing efficiency. Specifically for planning shortest paths in grid maps, the Field D* algorithm [Fer06b] has been developed. It can plan paths that lead through any point on the edge between two adjacent cells and is thus not restricted to the cell centers. This allows motions with arbitrary angles that are not restricted to grid-inherent 45° steps. This generally leads to smoother and hence shorter geometric paths through a grid map. A different approach is used by the Adaptive A* algorithm [Koe06]. It incrementally improves the heuristic after each planning cycle by exploiting information from the optimal solution of the previous planning cycle. This concept is further extended by the Multipath Adaptive A* (MPAA*) [Her14], which can additionally reuse parts of the previous solution that are still valid despite the changed environment. A similar replanning strategy is also known from sampling-based motion planning algorithms like the Multipartite Rapidly-exploring Random Tree (MP-RRT) [Zuc07].

Replanning algorithms usually perform a backward search from the goal to the start because it is assumed that the goal is static and that the robot is moving toward the goal. As a result, only the last part of the search graph changes, which benefits the efficient incremental update of the solution. For the intended application in this thesis, incremental replanning algorithms have some severe disadvantages: Due to the backward search, it is difficult to consider dynamic obstacles, since their prediction is based on the elapsed time from the start state, which is not known until finally reaching the start state during the backward search. The second drawback is the fact that, depending on the extent of changes in the environment, the effort to repair a previous solution may exceed the effort for planning from scratch. Thus, incremental replanning algorithms are usually capable of reducing the *average* planning time, but, due to the overhead for bookkeeping, they tend to increase the *worst case* planning time. The latter is, however, of particular importance in the presence of dynamic obstacles since a timely availability of a valid obstacle avoidance maneuver is essential for effective collision avoidance.

Anytime algorithms

Anytime algorithms constitute the second class of planning algorithms. They focus on a quick initial solution, which may be suboptimal. This solution is then iteratively refined until the maximum allotted computation time is exceeded for the current planning cycle or the optimal solution is found. The simplest approach is the utilization of A* with an increased heuristic inflation factor $\epsilon > 1$, which focuses the search toward the goal [Poh70]. This method is called Weighted A* and usually results in a faster, albeit possibly suboptimal, solution. If there is computation time left for the current planning cycle, the search is repeated with successively reduced ϵ until finally the optimal solution is found for $\epsilon = 1$ or, alternatively, the time budget is exhausted. Planning each refinement step from scratch, however, is unnecessarily expensive. This is why Anytime Repairing A* (ARA*) has been proposed [Lik03a], which is capable of reusing information from previous iterations (for details see Section 5.2.2). Once again, the algorithm is based on the successive reduction of the heuristic inflation factor ϵ until the optimal solution is found.

The rapid availability of a valid, i.e., collision-free, initial solution is particularly important for motion planning in the presence of dynamic obstacles. Even if this initial solution may be suboptimal, it is still a valid solution and thus avoids collisions effectively. If the time budget allows, the subsequent refinement steps can still provide the robot with the optimal solution. Besides the successive reduction of ϵ, the *anytime* strategy can be further extended to return only partial solutions: The authors of [Piv09a] start the search with an initially large goal region and then iteratively reduce the size of the goal region whenever a state in the goal is about to be expanded and if there is still some computation time left.

Anytime replanning algorithms

The third class of graph search algorithms contains the anytime replanning algorithms, which combine the advantages of incremental replanning and anytime algorithms. They are able to improve an initially suboptimal solution in an anytime fashion and in addition can exploit information from previous planning cycles. A prominent member of this class of algorithms is Anytime Dynamic A* (AD*) [Lik05], which is compared to ARA* in [Lik08]. The flexibility of anytime replanning algorithms is accompanied by an increased implementation complexity and besides the advantages they also inherit the disadvantages of incremental replanning algorithms, namely the backward search from goal to start and the increased worst case complexity if the environment has changed substantially. While the latter is unlikely for on-road driving scenarios, unstructured environments are more challenging in this respect. Consider for example a robot that drives around the corner of a building: Suddenly a previously unknown wall may come into view and thus change a large part of the occupancy grid map.

The fact that almost all of the teams of the DARPA Urban Challenge used A* or one of the above variants for planning in unstructured parking areas [Dol10] shows that the application of heuristic graph search algorithms for mobile robot motion planning has proved its worth in practice and forms the backbone of virtually all state-of-the-art mobile robot motion planning algorithms.

2.7 Planning with Graduated Fidelity

Robot motion planning is computationally demanding because of the high-dimensional search space commonly involved. This is why today most research concentrates on making existing methods more efficient.

A multi-resolution path planner for relatively uncluttered high-dimensional configuration spaces was proposed in [Boh01] for holonomic robots. The algorithm iteratively refines an implicit grid until a collision-free path is found or the final discretization size is reached.

Other methods exploit the fact that the requirements on planning accuracy decrease with increasing distance from the robot. A combination of a local trajectory generator, which simulates the vehicle's dynamics for a certain distance, and a simple global grid planner, which plans the remaining path to the goal, was proposed in [Kel06]. This two-staged concept was extended in [Kus09] to use a time-parametrized state lattice planner for the local planning step. However, both methods are not able to find kinematically feasible paths up to the goal because they only considered a simple grid for the global planning.

Besides the combination of local and global planners, multi-resolution state lattice methods have been developed, which apply a fine resolution near the start and goal, and a coarse resolution in between [Lik09; Ruf09a]. The term *graduated fidelity* was introduced in [Piv08] to describe a planning strategy that may vary in resolution and dimensionality. The authors proposed a planner that uses a four-dimensional (position, heading, and curvature) state lattice in the vicinity of the vehicle and an eight-connected grid (position only) elsewhere. Similar to the above-mentioned algorithms, this generally leads to kinematically infeasible paths. Furthermore, their proposed algorithm does not consider time-parametrized planning and is thus inapplicable for true motion planning among dynamic obstacles. To mitigate this problem, the authors propose a replanning approach in order to increase the planning frequency [Piv09b].

Path planning with adaptive dimensionality of the configuration space was also applied in the context of robotic manipulation [Coh11]. The approach is an extension to manipulation planning based on motion primitives [Coh10]. Depending on the distance to the goal, motion primitives with variable dimensionality are employed during a graph search, which allows for precise end effector positioning while still providing good overall planning performance.

The method suggested in [Goc11] exploits a hybrid-dimensional search graph: The algorithm first plans a path with hybrid dimensionality and then tries to track that path using a high-dimensional planning. The search space dimensionality is adapted such that difficult regions are permanently represented using high-dimensional states. The method was extended in [Zha12] for application in changing environments. For this purpose, a three-dimensional search space (position and heading) was used for the high-dimensional planning and a simple two-dimensional search space (position only) for the low-dimensional states. The method was applied to planning for mobile manipulation in [Goc12] and extended to incorporate an incremental planning scheme in [Goc13]. These methods are pure path planning methods and do not consider vehicle dynamics or dynamic obstacles.

2.8 Unresolved Issues

The basic path planning problem (as defined in Section 1.1.1) can be considered solved thanks to the massive amount of research carried out in the last decades [LaV11b]. However, especially for mobile robot applications, the strong focus on configuration space planning is insufficient. The nonholonomic nature of most mobile robots imposes complex restrictions on the possible motions. This is in contrast to the case of well-known industrial manipulator kinematics and requires the explicit incorporation of differential constraints during the planning. In addition, applications exhibiting fast dynamics usually require to plan high-dimensional trajectories consisting of all relevant state variables. But for dynamic environments, even this is still inadequate. In the presence of dynamic obstacles, these obstacles have to be taken into account explicitly during the trajectory planning. For this purpose, a temporal dimension needs to be added to the search space so that the motion planning can check for collisions with the predicted obstacles. All these requirements result in a high-dimensional planning problem with challenging demands on planning responsiveness for safe real-time operation. The common solution for this problem is the separation of the motion planning problem into several consecutive steps, e.g., a coarse global path planning and a subsequent local trajectory generation in order to track the global path and avoid collisions (see Section 2.3). This separation leads to suboptimal results by design due to different optimization scopes for global and local planning.

To this day, there exists no holistic resolution-optimal approach that combines global path planning over long distances and multiple waypoints with reactive local trajectory planning among moving obstacles in a unified, consistent, and convenient way.

To quote Steven M. LaValle [LaV11b]:

> [D]ifferential constraints, feedback, optimality, sensing uncertainty, and numerous other issues continue to bring exciting new challenges. In some sense, combining the components [...] leads to merging planning and control theory. Thus, the subject of planning at this level might just as well be considered as algorithmic control theory in which control approaches are enhanced to take advantage of geometric data structures, sampling-based searching methods, collision-detection algorithms, and other tools familiar to motion planning. The wild frontiers are open, and there are plenty of interesting places to explore.

Chapter 3

Multi-Resolution State Lattices with Hybrid Dimensionality

As was pointed out in Section 2.8, the limited integration of global and local planning in today's algorithms is a big issue on the way to powerful mobile robot navigation methods. This thesis presents an approach that tackles the problem of separated local motion and global path planning by introducing a unified, consistent method for mobile robot motion planning in unstructured, dynamic environments. The method builds on the concept of planning in state lattices that was described in-depth by Pivtoraiko, Knepper, and Kelly in [Piv09c].

There are basically two possibilities for search-based motion planning algorithms: One class (e.g., Hybrid A*) obtains the discrete search space by directly discretizing the input space of the robot. This generally leads to a set of reachable states that forms a tree (see Figure 3.1). The other class of

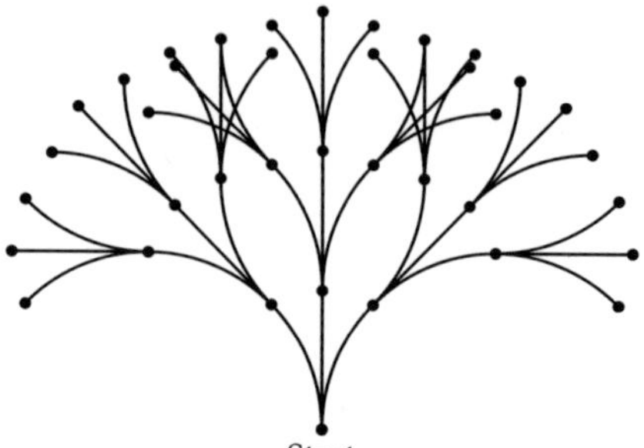

Figure 3.1: Uniform input sampling: the reachable set forms a tree.

algorithms directly discretizes the state space and subsequently constructs robot controls that move the dynamical system between those discrete states; thus, the resulting reachable set forms a lattice (see Figure 3.2). The significant advantage of this lattice structure is that it reduces the branching factor of the associated search graph because the number of states in a certain region of the state space is bounded. Thus, different path hypotheses might be merged again during the graph search. Details on the state of the art regarding planning in state lattices can be found in Section 2.6.2.

Planning trajectories to distant goals is computationally expensive. Therefore, current algorithms either consider only a very short time horizon for time-parametrized motion planning and then fall back to a kinematically infeasible grid search [Kus09], or they operate in a confined search space only, e.g., in on-road-only planning scenarios [Zie09]. The method proposed in this thesis extends the ordinary state lattice planning scheme with a search space representation consisting of variable resolution and dimensionality. This reflects the idea of decreasing relevance of certain state variables with increasing distance and time from the robot's current state.

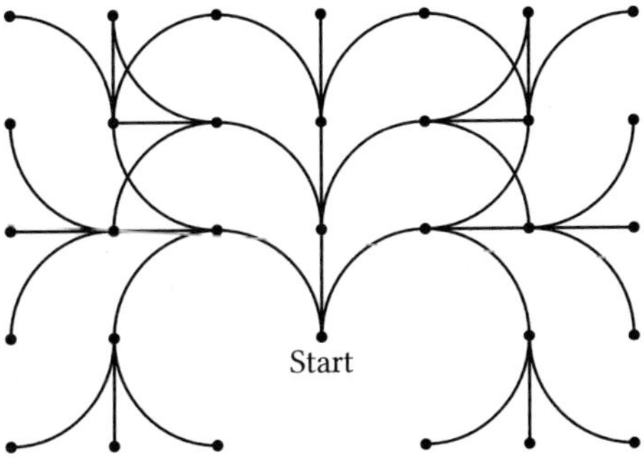

Figure 3.2: Uniform state sampling: the reachable set forms a lattice.

3.1 Overview of the Planning Concept

This section gives a short overview on the concept of the motion planning algorithm proposed in this thesis in order to provide some guidance on the relevance of the following sections in the overall context. The planning algorithm can be roughly split into four major parts.

Construction of discrete search space

In a first step, the transition from a continuous state space to a discrete representation of the search space is made. This discrete representation is called a state lattice, which gave this class of motion planning algorithms its name. Sections 3.2 to 3.4 describe the procedure for obtaining this discrete representation from a continuous state space.

Sampling of motion primitives

Next, a method to move the dynamical system along the quantized states is developed. For this purpose, motion primitives are employed that are guaranteed to start and arrive at quantized states only while still correctly capturing the robot's dynamics. The generation of such motion primitives is explained in detail in Section 3.5.

Introduction of varying resolution and dimensionality

Near the robot, time-parametrized planning is performed with maximum fidelity in order to incorporate the robot's dynamics and dynamic obstacles thanks to an explicit representation of time for the imminent future. A systematic gradual fidelity reduction (in terms of both dimensionality and resolution) allows for fast planning speed while maintaining kinematic feasibility throughout the whole solution (see Sections 3.6 and 3.7). This is even true for far-distant regions, where alternative algorithms often fall back to a mere grid search.

Search in the lattice

By concatenating the motion primitives, the motion planning problem can finally be reduced to an ordinary search for the shortest path in a graph, for which a variety of well-known algorithms exists. Particular attention is payed to the seamless integration of regions with different fidelity into *one* consistent graph. The proposed planning algorithm produces hybrid solutions being partly trajectory and partly path (see Chapter 5).

3.2 State Lattices

A lattice Λ is a discrete subset of a vector space and the generalized concept of a grid. For the vector space $\mathbb{R}^n$, a lattice can be specified as (see [Buh08])

$$\Lambda = \left\{ \sum_{i=1}^{m} k_i \mathbf{b}_i \,\middle|\, k_i \in \mathbb{Z} \right\}, \quad m \leq n \tag{3.1}$$

and thus consists of regularly spread points defined by all possible integer combinations of m linearly independent basis vectors $\mathbf{b}_i \in \mathbb{R}^n$. It was proven in [Bic02] that the reachable set of a dynamical system represented in chained form can be constrained to a lattice if the set of admissible controls is chosen appropriately. However, the practical implications of this property are rather limited. In the context of mobile robots, it was shown in [Ruf10] that the necessary transformation of the system equation to its chained form representation induces an unfavorable irregularity of the quantization of some states, and to make matters even worse, the quantization of some states becomes configuration-dependent, e.g., the steering angle quantization becomes dependent on the heading.

In order to achieve a more evenly distributed discretization, the following notion of a state lattice is less rigorous compared to the strict mathematical formulation (3.1) of a lattice. In this thesis, the term *state lattice* is used to describe a discrete subset of the robot's state space $\mathcal{S}$. A state lattice L covers the whole state space and can be thought of as some kind of state space quantization. It is possible to define irregular custom quantization values for certain dimensions of $\mathcal{S}$ (e.g., for heading or velocity); however, it might be advantageous to use a regular pattern, whenever possible, for a more concise representation of feasible motions (see Section 3.5).

A state lattice is constructed by discretizing the set of admissible values for each state variable s_i of the state vector[1] $\mathbf{s} \in \mathcal{S}$ individually and subsequently combining the respective sets. Let S_i be the quantized set of admissible values for the state variable s_i. The quantization can be performed in multiple ways. The straightforward approach is to use a regular discretization scheme

$$S_i := \{\, k\delta_{s_i} \mid k \in \mathbb{Z} \,\} \tag{3.2}$$

1. In the remainder of this thesis, the state vector $\mathbf{s} = [s_1\ s_2\ \ldots]^\top$ will often be more generally considered as a tuple and thus written $\mathbf{s} = (s_1, s_2, \ldots)$.

with δ_{s_i} being the quantization step size for state variable s_i; thus, S_i itself is a one-dimensional lattice in the mathematical sense. However, any other quantization scheme is also possible. The most general discretization approach is the explicit enumeration of quantization points $\tilde{s}_{i,k}$ for a state variable s_i:

$$S_i := \{ \tilde{s}_{i,1}, \tilde{s}_{i,2}, \tilde{s}_{i,3}, \dots \} . \tag{3.3}$$

After discretizing each of the n dimensions of the state space $\mathcal{S}$ individually, the state lattice L, embedded in $\mathcal{S}$, is defined by the Cartesian product of all one-dimensional sets S_i of admissible values:

$$L := \prod_{i=1}^{n} S_i . \tag{3.4}$$

The quantized states constituting the state lattice are called *lattice points*

$$\tilde{\mathbf{s}} \in L \subset \mathcal{S} . \tag{3.5}$$

They are decorated with a tilde to disambiguate them from their continuous counterparts.

For the application of a discrete search space to an actual (i.e., continuous) planning problem, it is necessary to develop means for the transition from continuous states to discrete states. Therefore, in order to find the nearest lattice point $\tilde{\mathbf{s}}$ for a continuous state $\mathbf{s}$, the mapping

$$\begin{aligned} \boldsymbol{\lambda} : \mathcal{S} &\to L \\ \mathbf{s} &\mapsto \arg\min_{\tilde{\mathbf{s}} \in L} \mu(\mathbf{s}, \tilde{\mathbf{s}}) \end{aligned} \tag{3.6}$$

is introduced. Thus, the nearest quantized state is obtained by minimizing its distance to the continuous state based on a metric

$$\mu : \mathcal{S} \times \mathcal{S} \to \mathbb{R} \tag{3.7}$$

that appropriately reflects the topology of the search space. An example is given in Section 3.4.

Utilizing the metric (3.7), the quantization error e_q can be defined as follows:

$$\begin{aligned} e_\text{q} : \mathcal{S} &\to \mathbb{R} \\ \mathbf{s} &\mapsto \mu(\mathbf{s}, \boldsymbol{\lambda}(\mathbf{s}))\,. \end{aligned} \tag{3.8}$$

It quantifies how well the continuous state $\mathbf{s}$ can be represented by a discrete state from L, i.e., the quantization error can be considered to be the generalized distance from the state $\mathbf{s}$ to its nearest discrete state $\tilde{\mathbf{s}} = \boldsymbol{\lambda}(\mathbf{s})$ that is part of the state lattice.

Finally, a trajectory that connects the quantized start state

$$\tilde{\mathbf{s}}_\text{s} = \boldsymbol{\lambda}(\mathbf{s}_\text{s}) \tag{3.9}$$

with the discrete goal region

$$\tilde{\mathcal{S}}_g = \left\{ \boldsymbol{\lambda}(\mathbf{s}) \,\middle|\, \mathbf{s} \in \mathcal{S}_g \right\} \tag{3.10}$$

can be specified as a sequence

$$(\tilde{\mathbf{s}}_k) = (\tilde{\mathbf{s}}_0, \tilde{\mathbf{s}}_1, \ldots, \tilde{\mathbf{s}}_K)\,, \quad \tilde{\mathbf{s}}_k \in L, \quad k = 0, \ldots, K \tag{3.11}$$

with $\tilde{\mathbf{s}}_0 = \tilde{\mathbf{s}}_\text{s}$ and $\tilde{\mathbf{s}}_K \in \tilde{\mathcal{S}}_g$. Two consecutive states $\tilde{\mathbf{s}}_k$ and $\tilde{\mathbf{s}}_{k+1}$ are linked in a particular manner using motion primitives in order to guarantee that the trajectory (3.11) correctly represents a feasible motion of the dynamical system. This is explained in depth in Section 3.5.

3.3 State × Time Lattices

The planning concept proposed in this thesis addresses path planning to possibly far goals as well as motion planning among moving obstacles in a dynamic environment. As was pointed out in Section 2.3, these two kinds of planning are commonly tackled separately. This is due to the large computational burden that results from the required dimensionality of the search spaces and the associated combinatorial complexity.

For planning in dynamic environments, it is at least necessary to include the temporal dimension in the search space so that the planning algorithm can check for collisions with predicted obstacles at any future point in time. If there exist further differential constraints for the robot's motion—as is the case with most mobile robot applications—more dimensions have to be introduced to properly reflect the robot's dynamics. Those additional dimensions are equivalent to time-derivatives of the robot's configuration and are already encoded in the robot's state space $\mathcal{S}$. The temporal dimension, however, has to be taken care of separately. For this purpose, the concept of state lattices is extended to the concept of *state × time lattices* in the following.

Without loss of generality, it can be assumed that the trajectory which is to be planned starts at an initial time $t_0 = 0$. The trajectory evolves with time strictly increasing, thus the set

$$\mathcal{T} := \{ t \in \mathbb{R} \mid t \geq 0 \} \tag{3.12}$$

denotes all future points in time (including the starting point $t_0 = 0$) for which a trajectory might be defined. Similar to the quantization of the individual state space dimensions, the temporal dimension has to be discretized in order to be incorporated into a search-based planning algorithm. Although it is possible to choose an arbitrary quantization scheme, a regular discretization simplifies the construction and handling of the motion primitive sets (see Section 3.5.3) without introducing any limitations on the planning problem. Therefore, the set of quantized points in time, T, is defined as follows throughout this thesis:

$$T := \{ k\delta_t \mid k \in \mathbb{N}_0 \} . \tag{3.13}$$

Analogously to the regular discretization scheme (3.2) for state variables, δ_t denotes the increment between two quantized points in time. The choice of δ_t naturally depends on the dynamics of the considered robotic system.

On the basis of the above definitions, the search space can be finally extended with a temporal dimension to form a state × time lattice. Since the resulting elements of the search space are given by all combinations of every quantized state with every quantized point in time, the state × time lattice L' is defined by

$$L' := \left(\prod_{i=1}^{n} S_i \right) \times T . \tag{3.14}$$

Thus, a state × time lattice is represented by the Cartesian product of the regular state lattice (3.4) and the set of quantized points in time (3.13).

The elements of the state × time lattice can be thought of as ordered pairs

$$(\tilde{\mathbf{s}}, \tilde{t}) \in L' \subset \mathcal{S} \times \mathcal{T} \tag{3.15}$$

with $\tilde{t} \in T$ denoting a quantized point in time. Similar to regular state lattices, it is necessary to provide a mapping of a continuous state $\mathbf{s}$ together with a continuous point in time t to the nearest point of the corresponding state × time lattice. Thus, by analogy to (3.6), the mapping

$$\begin{aligned} \boldsymbol{\lambda}' : \mathcal{S} \times \mathcal{T} &\to L' \\ (\mathbf{s}, t) &\mapsto \operatorname*{arg\,min}_{(\tilde{\mathbf{s}}, \tilde{t}) \in L'} \mu'(\mathbf{s}, t, \tilde{\mathbf{s}}, \tilde{t}) \end{aligned} \tag{3.16}$$

is introduced, which now also includes the temporal dimension in the corresponding metric

$$\mu' : \mathcal{S} \times \mathcal{T} \times \mathcal{S} \times \mathcal{T} \rightarrow \mathbb{R} . \tag{3.17}$$

As previously stated, the start time of the trajectory which is to be planned can be set to $t = 0$ without loss of generality. This results in a quantized start state-time pair

$$(\tilde{\mathbf{s}}_\mathrm{s}, 0) = \boldsymbol{\lambda}'(\mathbf{s}_\mathrm{s}, 0) . \tag{3.18}$$

If the time of arrival at the goal region is left unspecified as a free parameter, a trajectory that explicitly associates a discrete point in time with each quantized state can be written as the sequence

$$\left((\tilde{\mathbf{s}}, \tilde{t})_k\right) = \left((\tilde{\mathbf{s}}, \tilde{t})_0, (\tilde{\mathbf{s}}, \tilde{t})_1, \ldots, (\tilde{\mathbf{s}}, \tilde{t})_K\right), \quad (\tilde{\mathbf{s}}, \tilde{t})_k \in L' , \tag{3.19}$$

where $(\tilde{\mathbf{s}}, \tilde{t})_0 = (\tilde{\mathbf{s}}_\mathrm{s}, 0)$ and $(\tilde{\mathbf{s}}, \tilde{t})_K \in \tilde{\mathcal{S}}_\mathrm{g} \times \mathcal{T}$ with $\tilde{\mathcal{S}}_\mathrm{g}$ defined according to (3.10).

3.4 Case Study: Mobile Robot Motion Planning

The previous two sections described the theory for constructing state × time lattices regardless of a specific application. Planning in state × time lattices is a universal concept and not limited to a particular kind of dynamical systems. However, this thesis focuses on the development of a motion planning method for mobile robots. For the sake of clarity and to demonstrate the practical applicability of the proposed concept, the developed motion planning method will be illustrated by the application to an exemplary robotic platform throughout this thesis.

3.4.1 Robot Model

The motion planning algorithms proposed in this thesis have been devised in order to extend the autonomous driving capabilities of robotic platforms developed at Fraunhofer IOSB. The unmanned ground vehicle *IOSB.amp Q2* is a wheeled mobile robot with four-wheel steering. Details on the design, equipment, and capabilities of this robot are given in Section 6.4.

The robot's dynamical model, which is used for the actual motion planning, can be specified as a system of differential equations [Pet13c]:

$$\begin{aligned} \dot{x}(t) &= v(t)\cos\theta(t) \\ \dot{y}(t) &= v(t)\sin\theta(t) \\ \dot{\theta}(t) &= \kappa\, v(t)\tan\beta(t) \\ \dot{v}(t) &= a(t) \end{aligned} \tag{3.20}$$

This is an incarnation of the generic nonlinear system model (1.7) with state

$$\mathbf{s} = (x, y, \theta, v) \tag{3.21}$$

and input

$$\mathbf{u} = (a, \beta)\,. \tag{3.22}$$

The state variables x and y denote the robot's position, θ the orientation and v the scalar velocity in driving direction. The kinematical constant κ is twice the inverse of the wheelbase (see Figure 3.3). The system's inputs are given by the acceleration a in driving direction and the steering angle β. Of course, the system equations could be extended to include further derivatives like jerk or steering velocity. However, the chosen model suffices for the computation of smooth trajectories and allows for a straightforward exemplification of the proposed planning concept.

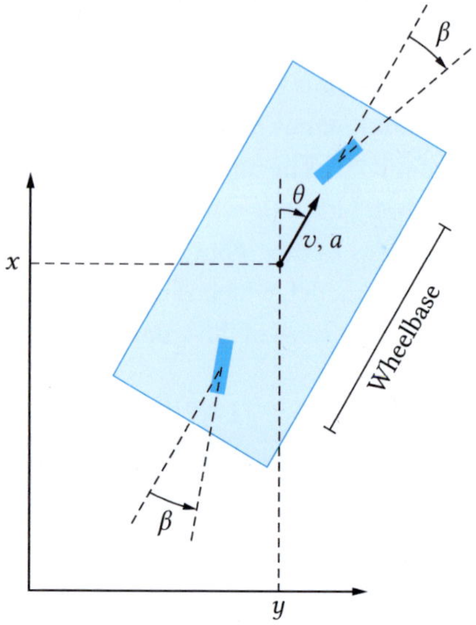

Figure 3.3: Instantaneous kinematics for a mobile robot with four-wheel steering. The geographical coordinate system used in this thesis follows the north-east-down (NED) convention, which is why the x-axis points upward and the y-axis to the right. Note that this is still a right-handed coordinate system.

3.4.2 Construction of State × Time Lattice

Using the exemplary motion model described in the previous section, this section shows how the construction of a state × time lattice is performed in practice. For this purpose, the structure of the state space corresponding to the system model (3.20) needs to be first analyzed. The associated state space $\mathcal{S}$ consists of the sets of admissible values for each state variable, which are discussed individually in the remainder of this section.

Position

The state variables x and y, which denote the position of the robot, are only limited by the size of the environment (or the range of the robot). Thus, for the state lattice construction, the position of the robot can be considered unbounded, i.e., $x \in \mathbb{R}$ and $y \in \mathbb{R}$.

Since—a priori—no position is preferred over another, it is reasonable to discretize the robot position in a uniform manner according to the general scheme (3.2). Thus, the sets of quantized x-positions X and quantized y-positions Y can be specified by

$$X := \left\{ k\delta_{xy} \,\middle|\, k \in \mathbb{Z} \right\}, \tag{3.23}$$

$$Y := \left\{ k\delta_{xy} \,\middle|\, k \in \mathbb{Z} \right\}. \tag{3.24}$$

Both x- and y-position use the same increment δ_{xy} because no direction is preferred over the other.

Heading

The robot's heading is represented by the angle θ. Its value is in the interval $[0, 2\pi)$ together with the additional identification of 0 and 2π. This representation is homeomorphic to the one-dimensional manifold

$$\mathbb{S}^1 = \left\{ (x, y) \in \mathbb{R}^2 \,\middle|\, x^2 + y^2 = 1 \right\}, \tag{3.25}$$

which is the unit circle, and which in turn is homeomorphic to $SO(2)$, the set of all 2D rotation matrices, called the *special orthogonal group* [LaV06]. The consideration of this particular topological structure is especially important for the definition of the metric (3.7), which needs to be defined for the state space (see Section 3.4.3).

At first glance, it may seem convenient to quantize the admissible heading values in a uniform fashion according to (3.2). Thus, the set of quantized

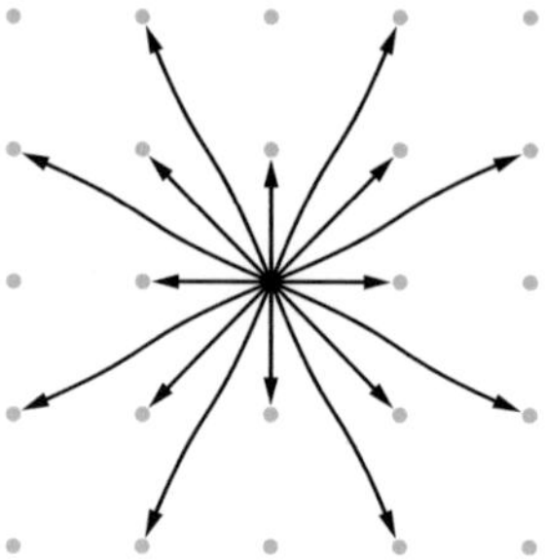

(a) Uniform quantization ($n_\theta = 16$).

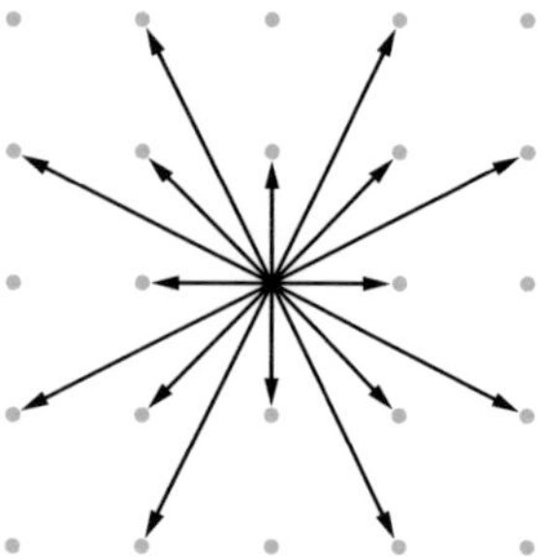

(b) Optimized for smooth motion.

Figure 3.4: Heading quantization strategies.

heading values would be given by

$$\Theta_{\text{uni}} := \left\{ i\frac{2\pi}{n_\theta} \;\middle|\; i \in \mathbb{N}_0 \wedge i < n_\theta \right\}, \quad \frac{n_\theta}{4} \in \mathbb{N}^+ . \tag{3.26}$$

The parameter n_θ denotes the total number of discrete headings. It is beneficial to specify it as a multiple of 4 in order to exploit the rotational symmetry of the $X \times Y$ sub-lattice with respect to the angle $2\pi/4$. However, the uniform quantization strategy has one major drawback: As soon as the number of discrete headings n_θ is chosen to be greater than 8, the robot may not be able to travel along points of the $X \times Y$ sub-lattice in straight motion. Instead, the robot needs to move along a wiggly line in order to reach a lattice point with a heading identical to the starting point (see Figure 3.4a).

Since discretizing the heading in a uniform way prevents straight movements, one should choose a discretization scheme that explicitly enables the robot to move along lattice points with constant heading as suggested in [Piv09c]. This can be achieved by defining the set of quantized heading values according to the non-uniform discretization scheme

$$\Theta_{\mathbb{Z}} := \left\{ \operatorname{atan2}(i,j) \,\middle|\, (i,j) \in \mathbb{Z} \times \mathbb{Z} \wedge (i,j) \neq (0,0) \right\} \tag{3.27}$$

with atan2 being the quadrant-aware arctangent function. Depending on the desired number of discrete heading values, the actual pairs (i, j) are restricted to a subset of $\mathbb{Z} \times \mathbb{Z}$ for practical applications. The example in Figure 3.4b shows this non-uniform heading quantization strategy for pairs $(i, j) \in \{ 0, \pm 1, \pm 2 \} \times \{ 0, \pm 1, \pm 2 \} \setminus \{ (0, 0) \}$, which results in a total number of $|\Theta_{\mathbb{Z}}| = 16$ unique discrete heading values. In this way, all sets $\Theta_{\mathbb{Z}}$ of discrete heading values obtained by (3.27) guarantee by design that a straight motion of the robot with constant heading $\tilde{\theta} \in \Theta_{\mathbb{Z}}$ leads through points of the $X \times Y$ sub-lattice. The remainder of this thesis assumes the use of this non-uniform discretization scheme, and therefore $\Theta_{\mathbb{Z}}$ will be briefly denoted by Θ.

Velocity

The robot's velocity v is limited by the physical capabilities of the robotic platform. Therefore, admissible values for the velocity are generally given by $v \in [v_{\min}, v_{\max}] \subset \mathbb{R}$. The interval may be directly inferred from the physical limitations of the platform or deliberately restricted to a subset of physically possible velocities. For example, the lower bound may be chosen to be $v_{\min} = 0$ in order to allow only forward motions in a particular scenario (e.g., driving on highways).

Various discretization schemes can be applied to obtain the set of quantized velocities V. The simplest approach is to use the regular discretization scheme (3.2) with the additional constraint $V \subset [v_{\min}, v_{\max}]$. However, it might be beneficial to explicitly enumerate the discrete velocity values following the scheme (3.3):

$$V := \{ \tilde{v}_1, \tilde{v}_2, \ldots \} \cup \{ 0 \} . \tag{3.28}$$

This allows for fine-grained planning in the anticipated range of velocities while considering velocities in less likely ranges with lesser sampling density. This reduces planning complexity considerably.

Whatever the chosen discretization scheme is, if the robot is operating in a dynamic environment, the robot has to be able to come to a stop in order to let other dynamic objects pass. This implies that $\tilde{v} = 0$ has to be included in the set of admissible discrete velocities (3.28).

Time

The quantization of time has already been discussed in detail in Section 3.3. Likewise, the exemplary state lattice employs the uniform quantization strategy (3.13), which is restated here for the sake of completeness:

$$T := \{ k\delta_t \mid k \in \mathbb{N}_0 \} . \tag{3.29}$$

Bringing it all together

The previous sections explained how the individual dimensions of the state × time space can be converted to their discretized counterparts. On this basis, the entire state × time lattice L' can finally be composed according to (3.14) by the Cartesian product of the respective discrete sets for each dimension:

$$L' = X \times Y \times \Theta \times V \times T . \tag{3.30}$$

This section has described the basic procedure for the construction of a state × time lattice using an example of a use case for mobile robot motion planning. This state × time lattice also constitutes the basis for the imple-

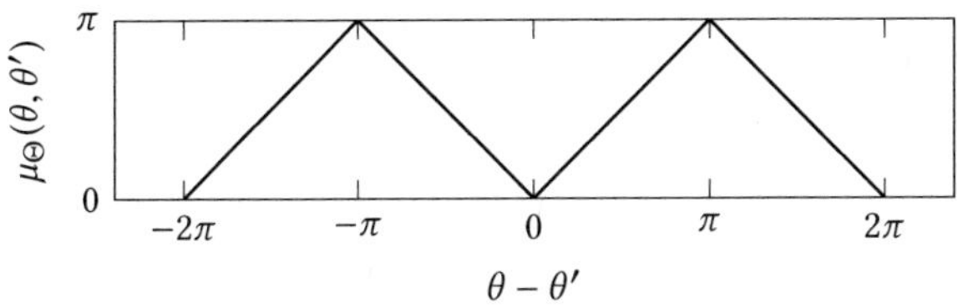

Figure 3.5: Metric for the set of robot headings

mentation and evaluation of the motion planning algorithm proposed in this thesis. The specific parameter values used for the discretization are described in Chapter 6 together with the corresponding results.

3.4.3 State Space Metric

It was mentioned in Section 3.3 that for each state × time space a corresponding metric has to be defined in order to transform the continuous planning problem into a discrete form. This metric depends on the topological structure of the planning space and this section explains how an appropriate metric can be defined for the state × time space in the mobile robot case study.

Special attention must be paid to the heading dimension, which is homeomorphic to the special orthogonal group $SO(2)$. A possible metric for $SO(2)$ is the traveled distance on the unit circle $\mathbb{S}^1$ between two angles. Because of the identification of the angles 0 and 2π, this metric can be defined by (see [LaV06])

$$\mu_\Theta(\theta, \theta') := \min\{|\theta - \theta'|, 2\pi - |\theta - \theta'|\} \tag{3.31}$$

and is depicted in Figure 3.5.

Furthermore, the following relation holds [LaV06]: Let $(X_i, \mu_{X_i}(x_i, x_i'))$, $i = 1, \ldots, n$ be n metric spaces consisting of the sets X_i with corresponding metrics $\mu_{X_i}(x_i, x_i')$. Moreover, let $Z = \prod_{i=1}^{n} X_i$ be the Cartesian product of the sets X_i. Then

$$\mu_Z(\mathbf{z}, \mathbf{z}') = \left(\sum_{i=1}^{n} \left(c_i \, \mu_{X_i}(x_i, x_i') \right)^p \right)^{1/p} \tag{3.32}$$

with $\mathbf{z} = (x_1, \ldots, x_n)$, $c_i > 0$, and $p \in \mathbb{N}^+$ is a metric for Z. The positive constants c_i can be chosen arbitrarily and function as scale factors for the respective sets X_i.

With (3.32), the metric (3.17) for the sample state × time space in the mobile robot case study can be composed of the metrics for each individual dimension. Thus, one possible metric is

$$\begin{aligned} \mu' : \mathcal{S} \times \mathcal{T} \times \mathcal{S} \times \mathcal{T} &\to \mathbb{R} \\ (\mathbf{s}, t, \mathbf{s}', t') &\mapsto \Big((c_x |x - x'|)^2 + (c_y |y - y'|)^2 \\ &\qquad + (c_\theta \, \mu_\Theta(\theta, \theta'))^2 + (c_v |v - v'|)^2 \\ &\qquad + (c_t |t - t'|)^2 \Big)^{1/2} \end{aligned} \tag{3.33}$$

with $\mathbf{s} = (x, y, \theta, v)$. In this metric, $\mu_\Theta(\theta, \theta')$ is defined according to (3.31) for the heading dimension. The remaining dimensions of $\mathcal{S} \times \mathcal{T}$ employ the standard Euclidean metric. The constants c_x, c_y, c_θ, c_v, and c_t are used to tune the ratio of the individual metrics, which is equivalent to scaling the respective dimensions. This is necessary because the metric (3.33) consolidates the different quantities represented by each dimension in a single expression. The state-space-only metric (3.7) can easily be derived from (3.33) by

$$\mu(\mathbf{s}, \mathbf{s}') := \mu'(\mathbf{s}, t, \mathbf{s}', t')\big|_{t=t'} \, . \tag{3.34}$$

3.5 Lattice-Conforming Motion

The previous section illustrated the methods to derive a discrete search space representation, namely a state × time lattice, from its continuous counterpart. In contrast to regular path planning approaches operating in $\mathcal{C}$-space only, the algorithm proposed in this thesis shall be able to obey the differential constraints of an arbitrary dynamical model (1.7). For this purpose, the following sections formulate the concept of motion primitives, which provide the basis for motion through the state × time lattice.

3.5.1 Motion Primitives

A motion primitive is a rule that moves the dynamical system from a time-augmented start state $(\tilde{\mathbf{s}}, \tilde{t}) \in L'$, located at a point of the state × time lattice, to a new time-augmented state $(\tilde{\mathbf{s}}', \tilde{t}') \in L'$, which remains a member of the state × time lattice. Let

$$\mathbf{s}(t) = \Phi(\mathbf{s}_0, t_0, \mathbf{u}, t), \quad t \geq t_0 \tag{3.35}$$

denote the evolution of the dynamical system (1.7) starting at state $\mathbf{s}_0$ and time t_0, with input $\mathbf{u}(\tau)$ applied until time t. The evolution can be defined by means of (1.9) as follows:

$$\Phi(\mathbf{s}_0, t_0, \mathbf{u}, t) := \mathbf{s}_0 + \int_{t_0}^{t} \mathbf{f}(\mathbf{s}(\tau), \mathbf{u}(\tau)) \, \mathrm{d}\tau \,. \tag{3.36}$$

For the system's evolution the following two properties hold:

1. After a zero-length duration, the dynamical system is still in its start state, i.e.,

$$\Phi(\mathbf{s}_0, t_0, \mathbf{u}, t_0) = \mathbf{s}_0 \,. \tag{3.37}$$

This proposition is immediately clear from the definition (3.36).

2. The evolution over a period $[t_0, t_2]$ can be calculated either at once or by concatenation, i.e.,

$$\Phi(\mathbf{s}_0, t_0, \mathbf{u}_1 + \mathbf{u}_2, t_2) = \Phi(\Phi(\mathbf{s}_0, t_0, \mathbf{u}_1, t_1), t_1, \mathbf{u}_2, t_2) \tag{3.38}$$

with $t_0 \leq t_1 \leq t_2$ and $\mathbf{u}_k(\tau) = 0$ for $\tau \in \mathbb{R} \setminus [t_{k-1}, t_k)$, i.e., the input $\mathbf{u}_k(\tau)$ is only active during the time interval $[t_{k-1}, t_k)$ and zero otherwise.

Proof. Substituting (3.36) into the left-hand side of (3.38) gives

$$\begin{aligned}
\Phi(\mathbf{s}_0, t_0, \mathbf{u}_1 + \mathbf{u}_2, t_2) &= \mathbf{s}_0 + \int_{t_0}^{t_2} \mathbf{f}(\mathbf{s}(\tau), \mathbf{u}_1(\tau) + \mathbf{u}_2(\tau))\,\mathrm{d}\tau \\
&= \underbrace{\mathbf{s}_0 + \int_{t_0}^{t_1} \mathbf{f}(\mathbf{s}(\tau), \mathbf{u}_1(\tau))\,\mathrm{d}\tau}_{\Phi(\mathbf{s}_0, t_0, \mathbf{u}_1, t_1) = \mathbf{s}(t_1)} + \int_{t_1}^{t_2} \mathbf{f}(\mathbf{s}(\tau), \mathbf{u}_2(\tau))\,\mathrm{d}\tau \\
&= \Phi(\Phi(\mathbf{s}_0, t_0, \mathbf{u}_1, t_1), t_1, \mathbf{u}_2, t_2)\,,
\end{aligned}$$

which is the right-hand side of (3.38). □

Hence, a state $\mathbf{s}(t_K)$ can be obtained by successively integrating over the intervals $[t_{k-1}, t_k]$ for $k = 1, \ldots, K$. The intermediate states $\mathbf{s}(t_k)$ evaluated during the integration can be collected into the sequence

$$\mathcal{T} := \left((\mathbf{s}, t)_k\right) = \left((\Phi(\mathbf{s}_0, t_0, \mathbf{u}, t_k), t_k)\right). \tag{3.39}$$

of state × time pairs, which represents the system's trajectory for the applied input $\mathbf{u}$.

A motion primitive is always associated with a particular start state $\tilde{\mathbf{s}}_0 \in L$. Throughout this thesis, it is assumed that the dynamical system of the robot, and hence the corresponding state evolution (3.36), is time-invariant, i.e.,

$$\Phi(\mathbf{s}_0, t_0, \mathbf{u}(\tau), t) = \Phi(\mathbf{s}_0, t_0 + \delta\tau, \mathbf{u}(\tau + \delta\tau), t + \delta\tau) \tag{3.40}$$

holds for all time delays $\delta\tau$. Thus, all motion primitives can safely be defined to start at time $\tilde{t}_0 = 0$ and can later be translated in time when needed.

Now a motion primitive m starting at $\tilde{\mathbf{s}}_0 \in L$ can be formally defined as the tuple

$$m := (\tilde{\mathbf{s}}_0, \tilde{\mathbf{s}}_e, \mathbf{u}, \mathcal{T}, \Delta t_m) \tag{3.41}$$

such that

$$\tilde{\mathbf{s}}_e = \boldsymbol{\lambda}(\Phi(\tilde{\mathbf{s}}_0, 0, \mathbf{u}, \Delta t_m)) \tag{3.42}$$

with Δt_m being the duration of the motion primitive, $\mathbf{u} : [0, \Delta t_m) \to \mathcal{U}$ being the system's input during the time interval $[0, \Delta t_m)$, and $\mathcal{T}$ being the trajectory (3.39) associated with the motion primitive. The state $\tilde{\mathbf{s}}_e$ denotes the discrete end state of the motion primitive and can be obtained by discretizing the last state of the trajectory $\mathcal{T}$. In order to be able to represent smooth motion of the dynamical system through the discrete lattice, the quantization error $e_q(\Phi(\tilde{\mathbf{s}}_0, 0, \mathbf{u}, \Delta t_m))$ of the end state needs to be as small as possible (and should ideally vanish). As a consequence, the construction of a motion primitive involves the solution of a boundary value problem (BVP) for the system equation (1.7). The generation of motion primitives is explained in detail in Section 3.5.4.

3.5.2 Motion Primitive Bunches

In general, there are many motion primitives all sharing the same start state $\tilde{\mathbf{s}}_0$ but having different duration and inputs and thus lead to different

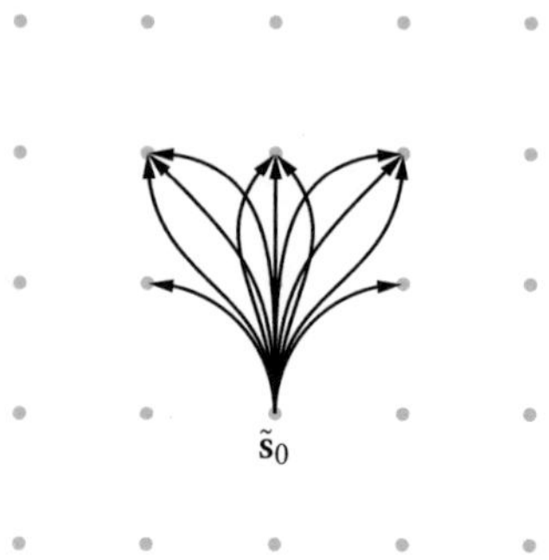

Figure 3.6: Sample motion primitive bunch for a simple 2D robot with eight discrete headings.

end states. These motion primitives are grouped in the set

$$\mathcal{B}(\tilde{\mathbf{s}}_0) := \{\, m \mid m \text{ starts at } \tilde{\mathbf{s}}_0 \,\}\,, \tag{3.43}$$

which will be called a *bunch* due to its characteristic shape (see Figure 3.6).

In order to be able to perform arbitrary motions based on motion primitive chaining, in theory each discrete state $\tilde{\mathbf{s}}$ requires a distinct corresponding motion primitive bunch. However, this is impossible in practice since the number of possible discrete states is generally unbounded. Fortunately, especially in mobile robotics, many dynamical models exhibit a form of translational invariance with respect to some state variables, which is the driving idea for the lattice representation of the state space. In this context *translational invariance* means that it does not matter whether a state variable is translated before or after the system's input is applied (cf. [Fra05]). In both ways, the dynamical system will be in the same state after integration. Formally speaking, the dynamical system is translation-invariant with respect to the state variable s_i if the relation

$$\Phi(\mathbf{s}_0 + \delta\mathbf{s}, t_0, \mathbf{u}, t) = \Phi(\mathbf{s}_0, t_0, \mathbf{u}, t) + \delta\mathbf{s} \tag{3.44}$$

holds for arbitrary values of δs_i, where δs_i is the i-th component of the translation vector $\delta\mathbf{s}$ and corresponds to the state variable s_i.

The dynamical model from the case study in Section 3.4 exhibits translational invariance with respect to the position variables x and y since $\mathbf{f}(\mathbf{s}, \mathbf{u})$, as defined in (3.20), does not depend on x or y, meaning the result of the integral in (3.36) is also independent of x and y. Thus, condition (3.44) holds for all $\delta\mathbf{s} = (\delta x, \delta y, 0, 0)$ with $\delta x, \delta y \in \mathbb{R}$. The translational invariance of the position state variables considerably reduces the number of motion primitive bunches that are necessary to represent all admissible motions. The sample application requires a motion primitive bunch only for every combination of discrete heading and velocity values. The starting position of the bunches can be safely set to be $x_0 = y_0 = 0$. The actual position of the robot can be computed by applying the motion primitive first (i.e., integrating the system model) and afterward adding the start offsets x_0 and y_0 to the system's state. Thus, in this example, a bunch $\mathcal{B}(\tilde{\mathbf{s}}_0)$ can be more concisely denoted by $\mathcal{B}(\tilde{\theta}_0, \tilde{v})$.

3.5.3 Motion Primitive Sets

The aggregate of all motion primitive bunches for all possible start states $\tilde{\mathbf{s}} \in L$ forms the *motion primitive set* $\mathcal{M}$ corresponding to the lattice L. A motion primitive set completely defines the admissible motion of the dynamical model through the corresponding state lattice and is constructed by

$$\mathcal{M} := \bigcup_{\tilde{\mathbf{s}} \in L} \mathcal{B}(\tilde{\mathbf{s}}) . \tag{3.45}$$

As mentioned in the previous section, it is not necessary to explicitly compute and maintain a motion primitive bunch for all potential start states if some state variables exhibit translational invariance. This applies to whole motion primitive sets, too. For example, the motion primitive set corresponding to the state lattice from the mobile robot sample application (see

Section 3.4) is more appropriately represented by

$$\mathcal{M} := \bigcup_{\substack{\tilde{\theta} \in \Theta \\ \tilde{v} \in V}} \mathcal{B}\left(\tilde{\theta}, \tilde{v}\right). \tag{3.46}$$

The total number of necessary motion primitive bunches, in order to fully define all feasible motions in this sample state lattice, is thus given by $|\Theta \times V| = |\Theta| \times |V|$.

3.5.4 Sampling of Motion Primitive Sets

The previous sections have introduced the notion of *motion primitives*, *motion primitive bunches* and *motion primitive sets* and explained how they can be used to define a lattice-conforming motion of the associated dynamical model. However, it has not been discussed yet how these motion primitives are actually constructed. This section describes the various possibilities for motion primitive construction and presents the method that was developed in the context of this thesis and implemented for the sample mobile robot application.

The sampling of motion primitives is an offline process, i.e., it is performed only once before the actual planning takes place. The resulting motion primitive set is then stored and exploited during the online planning phase. The shift of time-consuming computations from the planning phase to an upstream offline phase is a key part of the motivation for planning in state lattices.

Usually, motion primitives are constructed for one bunch at a time. According to the definition (3.41), constructing a motion primitive means finding an input $\mathbf{u}(t)$ which, when applied for the duration $\Delta t_{\mathrm{m}} \in T$, moves the dynamical system to a state $\tilde{\mathbf{s}}(\Delta t_{\mathrm{m}}) \in L$, which again is part of the state lattice. Generally, there is an infinite number of potential endpoints and thus motion primitives, so that a representative subset needs to be determined. A widely used approach is to directly choose the set of desired end states $\tilde{\mathbf{s}}(\Delta t_{\mathrm{m}}) \in L$ and then compute the corresponding inputs $\mathbf{u}(t)$ that move the dynamical system to these states. This constitutes a classic BVP, which usually needs to be solved by numerical methods.

The construction of motion primitives for planning in state lattices was first mentioned in Pivtoraiko's technical report [Piv04], which introduced the concept of motion planning in state lattices. The employed method, presented in detail in [Piv05a], uses curvature polynomials computed by inverse path generation based on the algorithm proposed in [Kel03]. It formulates the motion primitive construction as the solution of an optimal control problem (OCP) by using a variant of Newton's numerical optimization algorithm. However, the method produces only geometric motion primitives (i.e., paths) without velocity or time component. Nonetheless, this motion primitive construction method proved very efficient and has been the basis for more advanced techniques like construction of motion primitives suitable for kinodynamic planning [Piv11]. All of these techniques generally rely on the availability of a BVP solver for the differential equation describing the dynamical model of the robot. However, such a solver may be difficult to obtain for complex dynamical systems. An additional challenge is the a priori choice of desired end states. It is hard to tell beforehand which end states and hence motion primitives should be contained in a motion primitive bunch.

Optimal sampling of paths has been an active research field of its own in the past decade. There is an immanent trade-off between the total number of motion primitives (fewer motion primitives result in faster planning speed) and the exact representation of the system's dynamics (higher fidelity means better solution quality). The authors of [Gre07] proposed a method to minimize the dispersion of paths, i.e., maximize their diversity, at the cost of solution optimality. A similar approach, presented in [Bra08], maximizes path diversity based on the survival probability of paths among obstacles in a grid-world. The method was extended in [Eri09] to arbitrary environments. The authors also underline the difficulty to determine the appropriate number of necessary motion primitives by stating that it is not clear whether the determining factor is the dimension of the state space, the dimension of the control space, or a combination of both. The quality of path sets obtained by different sampling strategies was compared in [Kne09]. The authors found that random subsets of the full path set may perform much better and much worse compared to the respective sampling strategies. The actual performance is difficult to determine a priori: e.g., although most random subsets showed poor performance, the best random subset turned out to be superior to the Green-Kelly approach [Gre07].

Since sampling of optimal path sets has been comprehensively studied by many researchers, it is not the focus of this thesis. Instead, this thesis concentrates on methods to build an efficient search space representation for high performance motion planning. These methods are independent of the motion primitive sampling procedure. Nonetheless, the planning method proposed in this thesis requires the availability of appropriate motion primitive sets, too. The goal of this thesis is a highly generic planning method and this is why a motion primitive sampling approach has been developed that is capable of producing motion primitive sets for arbitrary dynamical systems.

Generic sampling approach

The algorithm developed in this thesis is based on probabilistic sampling of a large number of possible motion primitive candidates [Pet14]. The method applies forward integration only. Thus, it is not necessary to solve a BVP, which can be very demanding depending on the complexity of the dynamical system. The only prerequisite is the capability to compute the state evolution (3.36) for a given start state and system input. Each motion primitive bunch is constructed separately, which allows for efficient parallelization of the algorithm. The final motion primitive set $\mathcal{M}$ is obtained by consolidating all bunches in a common set according to (3.45).

The algorithm for sampling a bunch $\mathcal{B}(\tilde{\mathbf{s}}_0)$ consists of two phases: During the *exploration phase*, feasible motion primitives are added to the bunch. During the subsequent *consolidation phase* (and also already during the exploration phase), the previously sampled motion primitives are constantly optimized. The detailed steps of the algorithm are as follows:

1. Create an initially empty bunch $\mathcal{B}(\tilde{\mathbf{s}}_0) = \emptyset$, and choose the maximum duration of a motion primitive $\Delta t_{\mathrm{m,max}} = n_{\max}\delta_t$ with $n_{\max} \in \mathbb{N}^+$. Furthermore, specify the total number N_{total} of samples to be evaluated for this bunch as well as the number of samples in the exploration phase, N_{expl}, where $N_{\mathrm{expl}} < N_{\mathrm{total}}$. Now perform the following steps N_{total} times to simulate and evaluate the motion primitive candidates.

2. Initialize the trajectory $\mathcal{T}$ to the empty set and repeat the following steps for $k = 1, \ldots, n_{\max}$:

   (a) Sample a random input $\mathbf{u}_k : [0, \Delta t_{\mathrm{m,max}}) \to \mathcal{U}$ such that

$$\mathbf{u}_k(t) = \begin{cases} \mathbf{u}_{\mathrm{rand}} & \text{for } t \in [t_{k-1}, t_k)\,, \\ \mathbf{0} & \text{otherwise}\ , \end{cases} \tag{3.47}$$

   wherein $t_k = k\delta_t$ and $\mathbf{u}_{\mathrm{rand}}$ is a constant input randomly sampled from a uniform distribution representing the space of valid inputs $\mathcal{U}$. In this way, it is possible to effectively include control limits (e.g., limited steering angle) in the motion primitive construction process. To further improve the feasibility of the motion primitive, admissible inputs might be restricted to a subset of $\mathcal{U}$ if it is foreseeable that a particular input value will lead to a violation of state constraints in the future (e.g., accelerating while already driving at maximum speed).

(b) Compute the new end state of the dynamical system

$$\mathbf{s}_\mathrm{e}(k\delta_t) = \Phi(\tilde{\mathbf{s}}_0, 0, \mathbf{u}(t), k\delta_t) \tag{3.48}$$

with

$$\mathbf{u}(t) = \sum_{i=1}^{k} \mathbf{u}_i(t) \tag{3.49}$$

and append $\mathbf{s}_\mathrm{e}(k\delta_t)$ to the current trajectory $\mathcal{T}$. The computation of the state evolution in (3.48) can be efficiently performed by reusing the result of the previous iteration thanks to the concatenation property (3.38).

(c) Check whether the trajectory $\mathcal{T}$ from $\tilde{\mathbf{s}}_0$ to $\mathbf{s}_\mathrm{e}(k\delta_t)$ violates any static or dynamic state constraints. If yes, dismiss this motion primitive and start again beginning with step 2; otherwise compute the quantization error $e_\mathrm{q}(\mathbf{s}_\mathrm{e}(k\delta_t))$. If the quantization error is less than a threshold $e_{\mathrm{q,max}}$, i.e., the end state $\mathbf{s}_\mathrm{e}(k\delta_t)$ is sufficiently close to a lattice point, construct a new motion primitive $m = (\tilde{\mathbf{s}}_0, \tilde{\mathbf{s}}_\mathrm{e}, \mathbf{u}, \mathcal{T}, k\delta_t)$ with $\mathbf{u}(t)$ defined according to (3.49) and $\tilde{\mathbf{s}}_\mathrm{e} = \boldsymbol{\lambda}(\mathbf{s}_\mathrm{e}(k\delta_t))$, and check whether $\mathcal{B}(\tilde{\mathbf{s}}_0)$ already contains a motion primitive with the same discrete end state $\tilde{\mathbf{s}}_\mathrm{e}$:

   i. *Motion primitive with same end state already in bunch:* The quality of the newly sampled motion primitive m and the already existing motion primitive m' is assessed. For this purpose, the quantization loss

$$J_\mathrm{q}(m) := \left(e_\mathrm{q}\left(\mathbf{s}_\mathrm{e}(k\delta_t)\right)\right)^2 + \alpha c(m) \tag{3.50}$$

is computed for each motion primitive. The quantization loss is composed of the square of the already computed quantization error e_q and the cost $c(m)$ for executing the motion primitive. This cost may include, for example, the length of the trajectory corresponding to the motion primitive or the necessary actuating energy (see Section 5.3). The weighting factor $\alpha \geq 0$ is used to scale the cost $c(m)$ such that cost and squared quantization error are in the same order of magnitude. Thus, a reasonable value for α is the ratio of the average expected cost and the squared quantization error threshold.

If $J_\mathrm{q}(m) < J_\mathrm{q}(m')$, i.e., if the newly sampled motion primitive m is superior to the existing motion primitive m', m' is replaced with m.

ii. *No motion primitive with same end state in bunch:* If still in exploration phase (i.e., number of evaluated motion primitives less than or equal to N_expl), add the newly sampled motion primitive m to the bunch.

3. Terminate the algorithm when the total number of evaluated motion primitives exceeds N_total.

The algorithm's pseudocode is provided in Appendix B, Algorithm 1 on page 210. After the individual bunches have been sampled for all relevant start states $\tilde{\mathbf{s}}_0$, the complete motion primitive set $\mathcal{M}$ is finally composed according to (3.45). An exemplary implementation of the sampling algorithm for the mobile robot case study is described in the next section.

As already mentioned, the proposed algorithm relies on simulation, i.e., system model integration, only. For very simple models (see next section), this may be done in closed form. For more complicated models, one has to resort to numerical methods like Euler or Runge-Kutta methods. However, the proposed sampling algorithm is not restricted to system models for which an analytical representation is available. In fact, Φ may be implemented in terms of an arbitrary numerical simulation algorithm as long as it is capable of computing the system's evolution given a particular starting state and system input. It might even be possible to capture trajectories of the real system together with the corresponding controls for a given time and use this data to extract potential motion primitive candidates. Regardless of the specific model representation, the dynamical system can be assumed to be a black box, which makes the proposed algorithm very flexible and generic.

Mobile robot example

This section illustrates the sampling algorithm using the mobile robot example from Section 3.4. The system model (3.20) allows for an almost closed form computation of the state evolution (3.36). For a system input $\mathbf{u}(t) = (a(t), \beta(t))$ that is constant during each interval $[k\delta_t, (k+1)\delta_t)$, the differential equations (3.20) can be converted to the following time-discrete recurrence equations [Pet13c]:

$$x_{k+1} = x_k + \begin{cases} \dfrac{\sin\theta_{k+1} - \sin\theta_k}{\kappa \tan\beta} & \text{if } \beta \neq 0 \\ \left(\frac{1}{2}a\delta_t + v_k\right)\delta_t \cos\theta_k & \text{if } \beta = 0 \end{cases} \tag{3.51a}$$

$$y_{k+1} = y_k + \begin{cases} \dfrac{\cos\theta_k - \cos\theta_{k+1}}{\kappa \tan\beta} & \text{if } \beta \neq 0 \\ \left(\frac{1}{2}a\delta_t + v_k\right)\delta_t \sin\theta_k & \text{if } \beta = 0 \end{cases} \tag{3.51b}$$

$$\theta_{k+1} = \theta_k + \left(\tfrac{1}{2}a\delta_t + v_k\right)\kappa\delta_t \tan\beta \tag{3.51c}$$

$$v_{k+1} = v_k + a\delta_t \tag{3.51d}$$

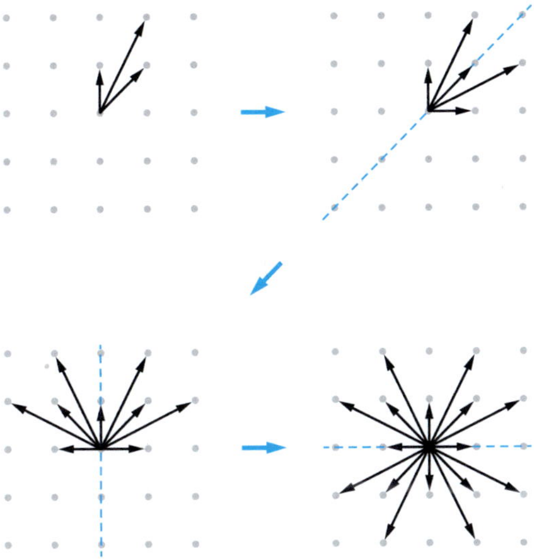

Figure 3.7: Construction of full motion primitive set through mirror symmetry of the first octant.

To keep notation short, the variables in this equation are named following the scheme $x_k = x(k\delta_t)$. With the set of recurrence equations (3.51), the state evolution (3.36) can be computed without expensive numerical integration.

As already mentioned in Section 3.5.2, the considered mobile robot system exhibits translational invariance with respect to the position variables x and y. Thus, only all combinations of heading and velocity values need to be considered in order to obtain the relevant start states for which motion primitive bunches are then computed.

Furthermore, the dynamical system features rotational as well as mirror symmetries. Thus, it is not necessary to sample a motion primitive bunch $\mathcal{B}\left(\tilde{\theta}, \tilde{v}\right)$ for every possible combination of $\tilde{\theta} \in \Theta$ and $\tilde{v} \in V$: Instead,

Table 3.1: Parameters for sampling of motion primitive set.

Parameter	Description	Value	Unit
δ_{xy}	Position increment	0.6	m
$\lvert\Theta\rvert$	Number of heading values	16	
V	Set of velocities	$\{-2, 0, 3\}$	m/s
δ_t	Time increment	0.5	s
$\Delta t_{\mathrm{m,max}}$	Max. motion primitive duration	2.0	s
N_{total}	Number of samples per bunch	1×10^8	
N_{expl}	Exploration threshold	5×10^7	
$e_{\mathrm{q,max}}$	Max. quantization error	0.2	
α	Weighting factor in (3.50)	0.002	

it is sufficient to consider start heading values in the first octant only, i.e., $\tilde{\theta} \in [0, \pi/4]$. Next, the remaining bunches are constructed by appropriate mirror symmetry (see Figure 3.7). Obviously, when applicable, this exploitation of system symmetries considerably reduces the construction time of the full motion primitive set $\mathcal{M}$.

Figure 3.8 shows an example of the development of the quantization error (3.8) and the quantization loss (3.50) as well as the total number of motion primitives per bunch during the sampling process. The sampling was performed using the parameters in Table 3.1 on an Intel Xeon E5-2687W CPU. Although the processing time was approximately 10 minutes, this is irrelevant because this offline motion primitive sampling is performed only once before the actual planning.

The computation of the quantization error e_{q} was conducted using a metric based on (3.33). The individual dimensions of the metric have been scaled such that the quantization error is normalized with respect to the quantization step sizes of each state space dimension. In addition, extra weight (factor 10) was given to the errors in x- and y-position to emphasize smooth motion along the grid. The cost $c(m)$ was computed by determining

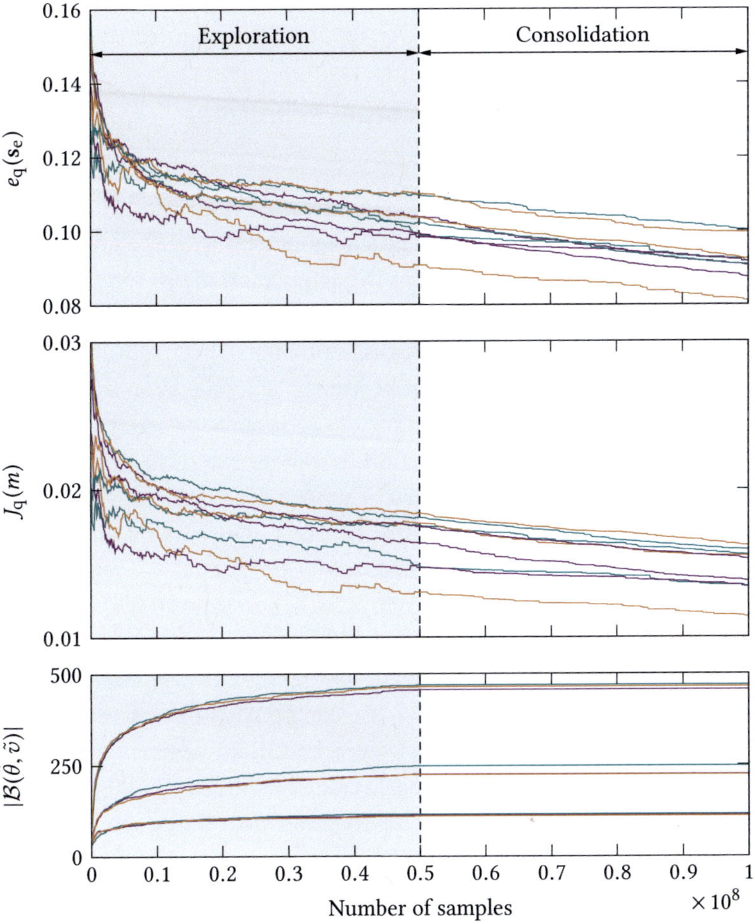

Figure 3.8: Sampling statistics: average quantization error (top), quantization loss (middle), and number of motion primitives per bunch (bottom) for a total of 1×10^8 samples.

the total path length of the motion primitive. The final motion primitive set contained a total of 13 916 motion primitives and the average number of motion primitives per bunch, $|\mathcal{M}|/|\Theta \times V|$, was approximately 290.

3.5.5 Decomposition of Motion Primitives

The motion primitive sampling performed according to Section 3.5.4 usually results in motion primitive bunches with a high number of motion primitives. This leads to a large branching factor during the graph search (see Chapter 5), which slows down the actual motion planning. Hence, it is desirable to reduce the number of motion primitives in $\mathcal{M}$ and thus the average number of motion primitives per bunch. Fortunately, the motion primitive set normally exhibits some kind of redundancy so that a subset of the full motion primitive set suffices to accurately encode the dynamics of the robot. The redundant motion primitives are those which can be reconstructed by the concatenation of shorter motion primitives and thus carry only limited additional information (see Figure 3.9 for an example). In this thesis, a motion primitive decomposition method was implemented that is closely based on the approach proposed in [Piv11]. It assumes that a small increase of execution costs could be tolerated if a motion primitive can be decomposed into a sequence of shorter motion primitives in return.

As was pointed out in [Piv11], the complexity of the decomposition would be exponential if all possible decompositions were enumerated explicitly. Therefore, a greedy approach, based on the fact that longer motion primitives are more likely to be decomposable than shorter motion primitives, is chosen. The algorithm works as follows.

First, all motion primitives $m \in \mathcal{M}$ are sorted by descending costs $c(m)$. The reduced set of atomic motion primitives $\mathcal{M}'$ is initialized to the empty set. Now, each motion primitive of $\mathcal{M}$ is checked for decomposability: In

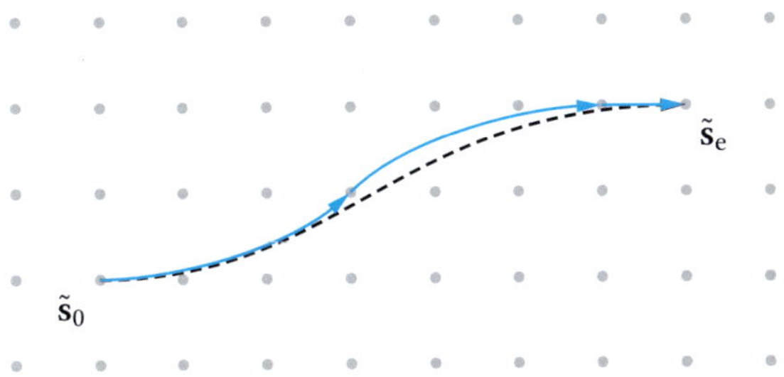

Figure 3.9: Decomposition of a motion primitive. The original motion primitive (dashed, black) is decomposed into three shorter motion primitives (cyan).

order to determine the next decomposition candidate, the motion primitive

$$m_{\max} = \arg\max_{m \in \mathcal{M}} c(m) \tag{3.52}$$

with maximum cost is selected and removed from $\mathcal{M}$. Next, an A* search [Har68] is applied to find the cost-optimal concatenation of motion primitives from $\mathcal{M} \setminus \{ m_{\max} \}$ which arrives at the same end state as the motion primitive $m_{\max}$. If no such sequence can be found or if the cost of the optimal sequence exceeds $c(m_{\max})$ by a certain factor ϵ_d, the motion primitive $m_{\max}$ is added to $\mathcal{M}'$ and the decomposition procedure continues with the next most expensive motion primitive according to (3.52) until all motion primitives have been considered for decomposition and thus $\mathcal{M}$ is empty and $\mathcal{M}'$ contains all remaining atomic motion primitives.

The cost factor $\epsilon_d > 1$ determines the relative increase in cost that is acceptable for decomposing a motion primitive in return. Reasonable values of ϵ_d are in the range 1 to 1.05, which equals a maximum accepted cost increase of 5 %. Of course, the exact value of ϵ_d is application specific and can be adjusted according to the user's preference. Thanks to the motion primitive decomposition, the cardinality of the motion primitive set and thus the average number of motion primitives per bunch, i.e., the branching factor during the graph search, could be reduced by approximately 70 % for the mobile robot example from Section 3.4.

Figures 3.10 to 3.12 show the resulting motion primitive bunches after applying the decomposition process to the set $\mathcal{M}$ from Section 3.5.4. Only the bunches in the first heading octant are shown, since the remaining bunches can be obtained by symmetry (see Section 3.5.4). Furthermore, only the projection of the motion primitives onto the xy-plane is visible from the plots; the actual motion primitives, of course, also contain a velocity and time component.

3.6 Hybrid Dimensionality

A motion primitive set that has been obtained by the procedure described in Section 3.5.4 allows for an accurate representation of the system's dynamics and thus forms the basis for high-fidelity motion planning. However, for practical applications, which require real-time operation, planning based on this complex representation is generally not feasible: The large number of motion primitives, which results from the high-dimensional search space (*curse of dimensionality*), implies a high branching factor during the graph search. Especially for distant goals, this may slow down the planning substantially. This trade-off between planning accuracy and planning speed is the major topic of this thesis. This section proposes an approach for speeding-up planning while still providing a high-quality planning result in relevant regions. The approach hinges on the idea that the required planning accuracy generally varies for different parts of the planning domain (see Figure 1.1). Especially for the near future, high planning quality is required as the imminent system inputs are derived from the planning result and passed to the robot. With increasing distance from the robot's actual state (both spatially and temporally), the accuracy requirements can be relaxed. This is, on the one hand, justified by the system and measure-

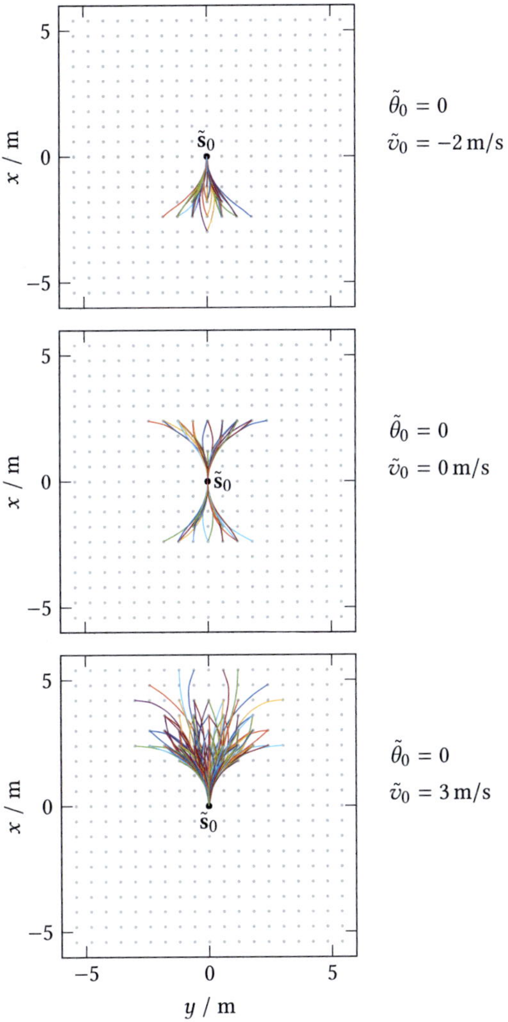

Figure 3.10: Motion primitive bunches for $\tilde{\theta}_0 = 0$.

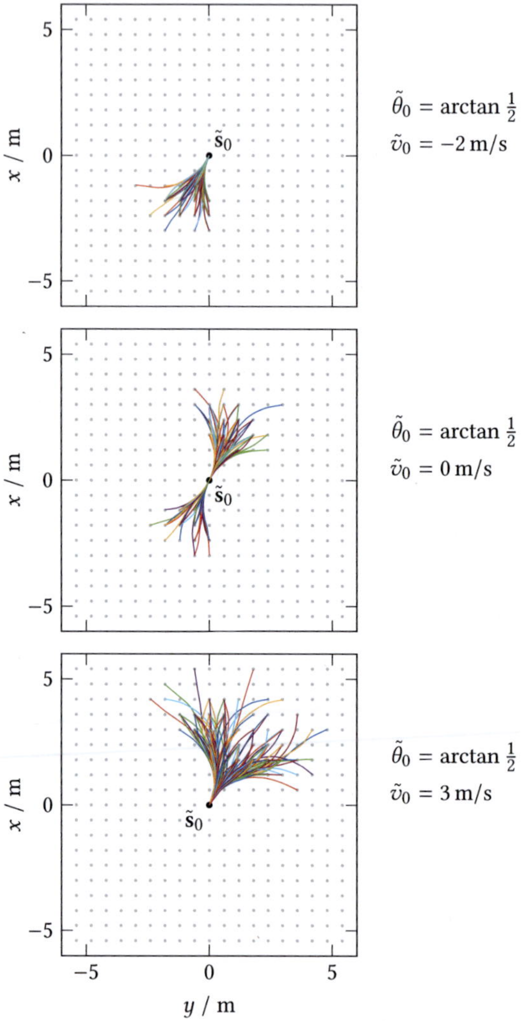

Figure 3.11: Motion primitive bunches for $\tilde{\theta}_0 = \arctan \frac{1}{2} \approx 26.6°$.

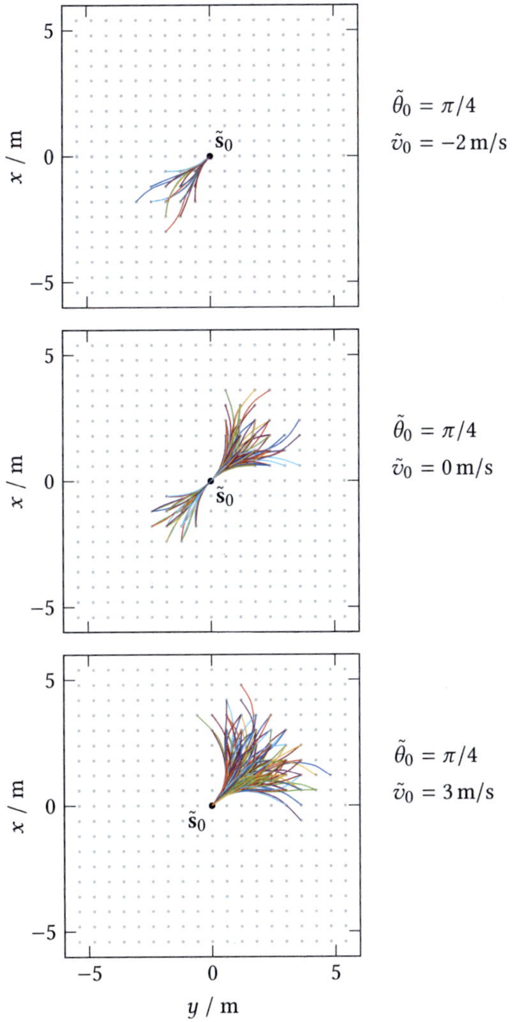

Figure 3.12: Motion primitive bunches for $\tilde{\theta}_0 = \pi/4 = 45°$.

ment uncertainties, which generally cause the robot to deviate from the initially computed plan after some given time. On the other hand, due to the dynamic environment and those uncertainties, plans are continuously recomputed anyway while the robot is moving toward the goal; thus, due to the receding horizon principle, each region along the traveled path eventually falls into the high-quality-planning region.

A viable option to reduce the computational complexity is the reduction of the dimensionality of the search space. In this thesis, an approach has been developed that allows for a systematic gradual dimensionality reduction for state × time lattices [Pet14; Pet13a]. Initial research on this topic was presented in [Piv08] for path planning with two different dimensionalities in the context of state-only lattices by combining kinodynamic path planning with a simple grid search, which, however, makes it impossible to consider the nonholonomic constraints of wheeled mobile robots.

For the formal derivation of the proposed gradual dimensionality reduction scheme, the notation introduced so far has to be slightly extended in order to make the involved dimensionality of some entities more explicit. The *dimensionality level* is expressed in terms of an additional superscript

$$d \in \{ 0, 1, \ldots, d_{\min} \}, \tag{3.53}$$

which denotes the number of applied dimensionality reductions. The level $d = 0$ corresponds to the maximum dimensionality used during the planning. Levels with less dimensions are labeled with indices in ascending order for decreasing dimensionality. The index $d_{\min}$ denotes the level with lowest dimensionality employed during the planning.

The variable dimensionality affects, first and foremost, the state (× time) lattice itself so that L^d denotes the state (× time) lattice that corresponds to the dimensionality level d. Maximum dimensionality (level $d = 0$) means planning in state × time space considering full robot dynamics; thus, $L^0 = L'$, i.e., L^0 is tantamount to the state × time lattice L' defined by (3.14). In general, the dimensionality of the state lattice L^d will briefly be denoted by $\dim L^d$.

Furthermore, all discrete states $\tilde{\mathbf{s}}^d$ are associated with a particular dimensionality level as they are elements of the corresponding state lattice L^d. The same applies for all symbols describing the motion through the lattice, e.g., the motion primitive bunch $\mathcal{B}^d$ and the motion primitive set $\mathcal{M}^d$. To keep notation short, the superscript d may be omitted if it is clear from the context. For example, $\mu^d(\mathbf{s}_1^d, \mathbf{s}_2^d)$ is a metric that, of course, matches the state's dimensionality level d and can thus be briefly denoted by $\mu(\mathbf{s}_1^d, \mathbf{s}_2^d)$.

3.6.1 Repeated Dimensionality Reduction

The dimensionality reduction is performed by successively projecting the state × time lattice onto subspaces with lower dimensionality. In what follows, regular states $\mathbf{s}$ and time-augmented states $(\mathbf{s}, t)$ will be consolidated in a *generalized state* $\mathbf{s}^d$ such that $\mathbf{s}^0 = (\mathbf{s}, t)$ by analogy to $L^0 = L'$. With this, the projection of generalized states can be concisely defined by

$$\begin{aligned} \pi^d : L^d &\rightarrow L^{d+1} \\ \tilde{\mathbf{s}}^d &\mapsto \tilde{\mathbf{s}}^{d+1} , \end{aligned} \tag{3.54}$$

where $\dim L^d > \dim L^{d+1}$. Once again, the superscript d may be omitted from π because the associated dimensionality level follows directly from the argument. In the simplest case, the projection (3.54) is defined to simply strip the appropriate dimensions, e.g., time $\tilde{t}$, from the state lattice such that

$$\pi\left(\tilde{\mathbf{s}}^0\right) = \pi\left((\tilde{\mathbf{s}}, \tilde{t})\right) = \tilde{\mathbf{s}} = \tilde{\mathbf{s}}^1 . \tag{3.55}$$

One possible choice of a projection scheme for the mobile robot example from Section 3.4 is as follows:

$$\tilde{\mathbf{s}}^0 = (\tilde{x}, \tilde{y}, \tilde{\theta}, \tilde{v}, \tilde{t}) \in L^0 = X \times Y \times \Theta \times V \times T \tag{3.56}$$

$$\pi(\tilde{\mathbf{s}}^0) = \tilde{\mathbf{s}}^1 = (\tilde{x}, \tilde{y}, \tilde{\theta}, \tilde{v}) \in L^1 = X \times Y \times \Theta \times V \tag{3.57}$$

$$\pi(\tilde{\mathbf{s}}^1) = \tilde{\mathbf{s}}^2 = (\tilde{x}, \tilde{y}, \tilde{\theta}) \in L^2 = X \times Y \times \Theta \tag{3.58}$$

In this equations, the various degrees of planning fidelity can clearly be seen: The planning in L^0—see (3.56)—corresponds to planning with full robot dynamics in a time-varying environment; planning in L^1—see (3.57)—yields a dynamically feasible trajectory (but now without incorporation of dynamic obstacles); and finally, planning in L^2—see (3.58)—can at least guarantee a kinematically feasible path (i.e., ignoring any time derivatives of the system's configuration variables). In theory, even a further reduction to $L^3 = X \times Y$ would be conceivable; however, this would make considering the nonholonomic constraints of a wheeled robotic platform during the motion planning impossible. Nonetheless, this might be of interest for practical applications if computation time is scarce. An approach that is based on short-term high-fidelity time-parametrized planning combined with subsequent kinematically infeasible planning in $X \times Y$ has been proposed in [Kus09]. This thesis, however, pursues a method that is able to gradually lessen the planning fidelity while guaranteeing a solution that is at least kinematically feasible for all segments up to the goal. This ensures that the robot is able to actually execute the planned motion.

In order to define the admissible motion through a lattice with reduced dimensionality, it is necessary to construct motion primitive sets with corresponding dimensionality. This is accomplished by individually projecting all high-dimensional motion primitives onto the respective subspace associated with the dimensionality level. Since in the end only the discrete start and end states of a motion primitive are relevant for the construction of the planning graph (see Section 5.1), it is sufficient to apply the dimensionality reduction only to those two components of the motion primitive. Hence, a motion primitive

$$m^d = \left(\tilde{\mathbf{s}}_0^d, \tilde{\mathbf{s}}_\mathrm{e}^d, \mathbf{u}, \mathcal{T}, \Delta t_\mathrm{m}\right), \tag{3.59}$$

belonging to the dimensionality level d, can be transformed by application of (3.54) to a motion primitive

$$m^{d+1} = \left(\pi(\tilde{\mathbf{s}}_0^d), \pi(\tilde{\mathbf{s}}_\mathrm{e}^d), \mathbf{u}, \mathcal{T}, \Delta t_\mathrm{m}\right) = \left(\tilde{\mathbf{s}}_0^{d+1}, \tilde{\mathbf{s}}_\mathrm{e}^{d+1}, \mathbf{u}, \mathcal{T}, \Delta t_\mathrm{m}\right) \tag{3.60}$$

for the new dimensionality level $d + 1$. Figure 3.13 shows this projection scheme in the context of the mobile robot case study.

For dimensionality levels with $d > 1$ the meaningfulness of the trajectory $\mathcal{T}$ with duration Δt_m and the system input $\mathbf{u}$ becomes somewhat limited because they are strongly linked to the high-dimensional start state $\tilde{\mathbf{s}}_0^0$. Nonetheless, valuable information can be extracted especially from $\mathcal{T}$, like the trajectory's geometry, which is required for collision avoidance (see Section 4.1).

By application of (3.43) and (3.45), motion primitive bunches

$$\mathcal{B}^{d+1}\left(\tilde{\mathbf{s}}_0^{d+1}\right) := \left\{ m^{d+1} \,\middle|\, m^{d+1} \text{ starts at } \tilde{\mathbf{s}}_0^{d+1} \right\} \tag{3.61}$$

and motion primitive sets

$$\mathcal{M}^{d+1} := \bigcup_{\tilde{\mathbf{s}}^{d+1} \in L^{d+1}} \mathcal{B}^{d+1}\left(\tilde{\mathbf{s}}^{d+1}\right) \tag{3.62}$$

for planning in L^{d+1} can be finally composed from the projected motion primitives (3.60).

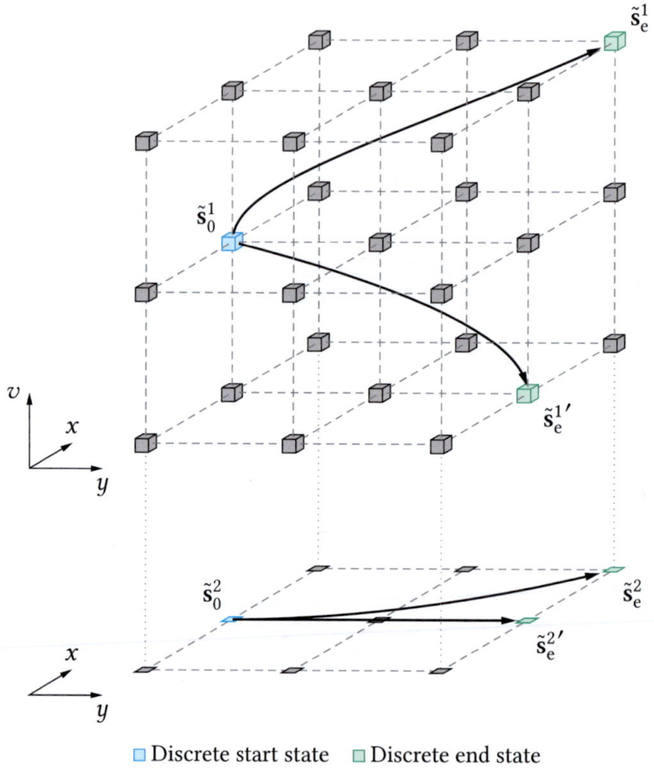

Figure 3.13: Projection of motion primitives from L^1 to L^2, see (3.57) and (3.58). The heading dimension is not shown.

3.6.2 Removing Redundant Motion Primitives

Due to the projection process, there will be multiple motion primitives with identical start and end states present in a bunch, i.e., for $d > 0$ there may exist two motion primitives

$$m^d = \left(\tilde{\mathbf{s}}_0^d, \tilde{\mathbf{s}}_e^d, \mathbf{u}, \mathcal{T}, \Delta t_m\right) \in \mathcal{B}^d\left(\tilde{\mathbf{s}}_0^d\right) \tag{3.63}$$

and

$$m^{d'} = \left(\tilde{\mathbf{s}}_0^{d'}, \tilde{\mathbf{s}}_e^{d'}, \mathbf{u}', \mathcal{T}', \Delta t_m{}'\right) \in \mathcal{B}^d\left(\tilde{\mathbf{s}}_0^{d'}\right) \tag{3.64}$$

with $\tilde{\mathbf{s}}_0^d = \tilde{\mathbf{s}}_0^{d'}$ and $\tilde{\mathbf{s}}_e^d = \tilde{\mathbf{s}}_e^{d'}$. This would result in unnecessary redundancy during the eventual graph search. However, the key intention of this thesis is the exploitation of repeated motion primitive projections in order to reduce the total number of motion primitives and thus the branching factor during the planning. For this purpose, only one motion is allowed to exist in a bunch $\mathcal{B}^d$ for any end state $\tilde{\mathbf{s}}_e^d$. This is ensured by the following procedure.

The projection is conducted bunch-wise. Whenever the new bunch of already projected motion primitives, $\mathcal{B}^d$, contains a motion primitive whose end state coincides with the end state of the currently projected motion primitive, a scoring of these two motion primitives is performed and the motion primitive with the higher cost is dropped. This scheme is analogous to the selection scheme for sampling the original, non-projected motion primitives (see Section 3.5.4). Thus, the same assessment rules, i.e., the quantization loss J_q, equation (3.50), may be employed. However, other scoring schemes may be used as well, e.g., it is sensible to put more weight on the execution costs of a motion primitive by using an increased value for the weighting factor α in (3.50) because the considered end states already exhibit small quantization errors due to the original sampling process.

In order to reduce the branching factor, i.e., the number of motion primitives per bunch, even further, the decomposition approach from Section 3.5.5 is also applied to each projected bunch. Thus, the algorithm attempts to replace long motion primitives by the concatenation of shorter motion primitives.

These two reduction strategies are able to significantly reduce the number of motion primitives per bunch. This allows for the construction of an efficient planning graph, which is the key motivation for the hybrid-dimensional planning algorithm developed during this thesis. The transition between different dimensionality levels during the graph search is described in Section 5.1.2.

3.7 Multiple Resolutions

The dimensionality reduction explained in the previous section is a valuable strategy for reducing the overall planning complexity. Moreover, the employed resolution is another determining factor to control the fidelity of the search space. Once again it is only the immediate future and environment that requires high planning quality; this accuracy requirement decreases with increasing distance (both spatially and temporally) from the actual robot position (cf. Figure 1.1). This approach is justified by the fact that measurement and system uncertainties make it necessary to continuously replan the robot's motion.

While the choice of the dimensionality level is independent of the environment (as long as $\mathcal{M}^{d_{\min}}$ correctly models the robot's kinematics), the choice of the resolution may directly affect the existence of a solution to the planning problem. For example, a solution involving a path through a narrow corridor may exist for planning with fine resolution but may be lost when switching to a more coarse resolution. Therefore, multi-resolution planning is a common strategy for mitigating these kind of problems. It was first mentioned in [Lik09] in the context of single-dimensionality state-only-based lattice planning. This thesis generalizes the concept and proposes the simultaneous utilization of different dimensionalities and regions with varying resolutions in order to find a compromise between completeness, solution quality and planning speed [Pet13b].

In order to integrate planning with variable resolution into the holistic motion planning concept, a *resolution level* is introduced by analogy to the dimensionality level from Section 3.6. The resolution level

$$r \in \{\, 0, 1, \ldots, r_{\min} \,\} \tag{3.65}$$

characterizes the quantization S_i^r of the individual state space dimensions. Here, the resolution level r has been attached as a superscript to the set of discrete state values S_i, in order to make this connection explicit. The level $r = 0$ denotes the finest resolution used during the planning. Levels with less refined resolution are labeled with indices in ascending order for decreasing resolution. The index $r_{\min}$ denotes the most coarse resolution employed during the planning.

For a state variable s_i that is discretized according to the regular scheme (3.2), the resolution level directly affects the quantization increment δ_{s_i}, which therefore becomes the resolution-dependent increment $\delta_{s_i}^r$. This is also true for state variables that are discretized in a custom way; however, in this case the procedure for obtaining the sets of discrete states S_i^r with different resolutions is slightly more complicated (see Section 3.7.1). Since the individual sets S_i^r eventually form the state ($\times$ time) lattice $L^{d,r}$, this lattice also becomes resolution-dependent and is thus henceforth carrying two superscripts to denote its associated dimensionality level d and resolution level r. From this follows directly that the discrete states $\tilde{\mathbf{s}}^{d,r}$, which essentially are the lattice points of $L^{d,r}$, are also resolution-dependent. The same is true for the resolution-dependent mapping $\boldsymbol{\lambda}^r$, see (3.16), which yields the nearest lattice point for a generalized[2] continuous state $\mathbf{s}^d$, and

2. For the notion of a *generalized state* see Section 3.6.1.

the associated quantization error e_{q}^{r}, see (3.8). Finally, as a consequence of the introduction of additional resolutions, the set of admissible motions through the lattice, which was defined in Section 3.5, needs to be additionally expressed in a resolution-dependent way and is therefore from here on denoted by $m^{d,r}$ for a motion primitive, $\mathcal{B}^{d,r}$ for a motion primitive bunch, and $\mathcal{M}^{d,r}$ for a motion primitive set.

3.7.1 Construction of Resolution-Specific Lattices

All lattices are inherently resolution-dependent due to their underlying discretization scheme. Thus, constructing multiple state (× time) lattices with various resolutions is similar to constructing individual lattices based on the procedure described in Sections 3.2 and 3.3 with appropriately chosen discretization strategies. However, special care must be taken while choosing the design parameters, i.e., the discretization scheme, of the lattice in order to guarantee smooth interoperability of lattices with different resolutions for the joint planning with multiple resolutions.

It is generally desirable that all solutions that can be found in a lattice L^{d,r_1} with a coarse resolution can also be found in a lattice L^{d,r_2} with a finer resolution ($r_1 > r_2$). This requires the set of admissible motions $\mathcal{M}^{d,r_1}$ to be also contained in $\mathcal{M}^{d,r_2}$, i.e.,

$$\mathcal{M}^{d,r_1} \subset \mathcal{M}^{d,r_2} \tag{3.66}$$

and

$$L^{d,r_1} \subset L^{d,r_2} \tag{3.67}$$

must hold for $r_1 > r_2$.

For the regular discretization according to (3.2), condition (3.67) can be easily satisfied by choosing the increments $\delta_{s_i}^{r}$ such that

$$\delta_{s_i}^{r_1} = n\,\delta_{s_i}^{r_2}, \quad n \in \mathbb{N}^{+} \tag{3.68}$$

for $r_1 > r_2$. From this follows immediately that $S_i^{r_1} \subseteq S_i^{r_2}$.

This trivial procedure for obtaining multiple-resolution representations is not applicable for dimensions of the state space $\mathcal{S}$ that are discretized by irregular discretization schemes, like the heading discretization scheme (3.27) or the velocity discretization scheme (3.28). Instead, the sets of discrete states S_i^{r+1} need to be directly obtained from their corresponding sets S_i^r with finer resolution. For an initial heading discretization that consists of 32 different orientations, i.e., $|\Theta^0| = 32$, this might be, for example, achieved by using only every other heading, resulting in $|\Theta^1| = 16$.

Even an entirely manual selection of the desired values is perfectly acceptable. This is, for example, useful to define the low-resolution set V^1 of discrete velocities by picking appropriate values from the high-resolution set V^0. It might even be sensible to choose $S_i^{r+1} = S_i^r$ for some of the dimensions of $\mathcal{S}$, but not for all dimensions simultaneously because then $L^{d,r_1} = L^{d,r_2}$ would hold, which would violate condition (3.67). In this case it would be meaningless to introduce an additional lattice since both lattices would be identical.

The sets of discrete states S_i^{r+1} and time T^{r+1} with reduced resolution form the new resolution-specific state × time lattice

$$L^{0,r+1} = \left(\prod_{i=1}^{n} S_i^{r+1} \right) \times T^{r+1} \,. \tag{3.69}$$

On the basis of this lattice the corresponding motion primitive sets $\mathcal{M}^{d,r+1}$ can be constructed using the method described in Section 3.5.4. Then, the projection procedure from Section 3.6 is applied to each resolution level individually in order to obtain state lattices with variable dimensionality.

While condition (3.67) is satisfied by design, extra measures need to be taken in order to satisfy condition (3.66). Let $\mathcal{M}^{d,r'}$ be a motion primitive set obtained by the motion primitive sampling procedure from Section 3.5.4.

The final motion primitive set $\mathcal{M}^{d,r}$ is determined by the following recurrence system:

$$\mathcal{M}^{d,r_{\min}} := \mathcal{M}^{d,r_{\min}}{}' \tag{3.70}$$

$$\mathcal{M}^{d,r-1} := \mathcal{M}^{d,r-1}{}' \cup \mathcal{M}^{d,r} . \tag{3.71}$$

This method guarantees that all motions that can be found in a low-resolution lattice can also be found in lattices with higher resolution and thus higher planning fidelity [Pet13b].

3.7.2 Application of Multiple Resolution Levels

To illustrate the multiple-resolution approach, Figure 3.14 shows two motion primitive bunches with different resolution levels. The bunch in Figure 3.14a is the high-resolution motion primitive bunch $\mathcal{B}^{1,0}(\tilde{\mathbf{s}}_0^{1,0})$ and the bunch in Figure 3.14b is the corresponding low-resolution motion primitive bunch $\mathcal{B}^{1,1}(\tilde{\mathbf{s}}_0^{1,1})$. Both bunches originate from the same start state, i.e., $\tilde{\mathbf{s}}_0^{1,0} = \tilde{\mathbf{s}}_0^{1,1}$. It can be seen that all the motion primitives from $\mathcal{B}^{1,1}$ are also contained in $\mathcal{B}^{1,0}$ in compliance with (3.66).

Beside its variable dimensionality, the final motion planning solution simultaneously comprises trajectory and path sections with different resolutions. The resolution level r and dimensionality level d can be chosen independently of each other. They constitute orthogonal concepts for varying the fidelity of the planning. Section 5.1.3 describes the transitions between search space regions with different resolution levels. The criteria for choosing a particular resolution level during the course of the planning are discussed in depth in Section 5.1.4.

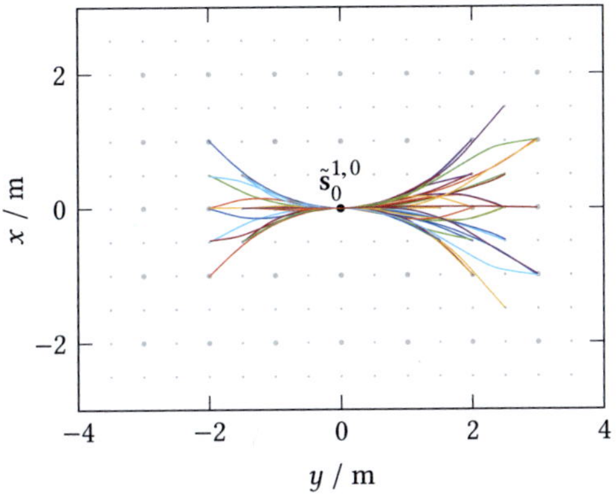

(a) High-resolution motion primitive bunch $\mathcal{B}^{1,0}$.

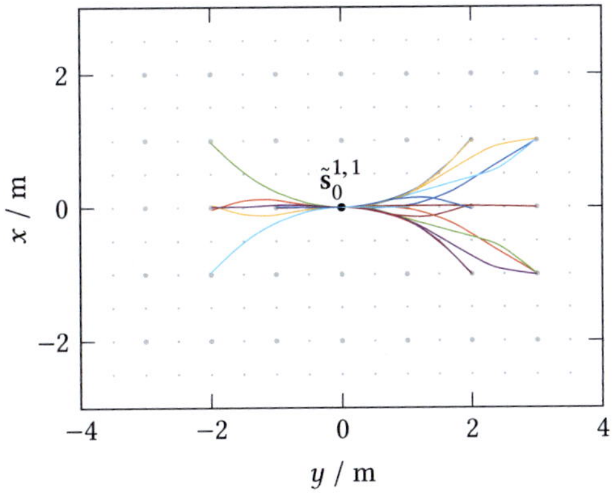

(b) Low-resolution motion primitive bunch $\mathcal{B}^{1,1}$.

Figure 3.14: Comparison of two bunches with different resolution level.

3.8 Multiple Waypoints

Moving along a chain of sub-goals g_i, $i = 1, \ldots, N$, is a common task in robotics. In the context of mobile robots, these sub-goals are also called *waypoints*. Each waypoint is associated with the set of states $\mathcal{S}_{g_i}$ which satisfy the desired (sub-)goal condition. This condition can be very restrictive (e.g., for a parking task with exact goal position and orientation) or rather open (e.g., a desired position with specified tolerance and arbitrary orientation).

For planning along multiple waypoints, the naive approach is to plan to the next waypoint only and then continue with a separate planning to the following waypoint. Especially if a goal configuration has many degrees of freedom, this naive approach may be problematic and will often lead to suboptimal global results. For example, consider Figure 3.15a: It is the robot's task to travel to the goal waypoint g_2 via waypoint g_1, which is located next to an obstacle. If the motions from the start to g_1 and from g_1 to g_2 are planned separately, a needless turning maneuver will result for the second segment because the robot's orientation when arriving at g_1 is disadvantageous for the subsequent motion to g_2. Hence, the overall solution will be suboptimal. In contrast to the naive planning approach, consider the combined planning for g_1 and g_2 in Figure 3.15b: The trajectory is directly planned to the final goal waypoint g_2 with the additional constraint that the trajectory must lead through the via-waypoint g_1, i.e., the trajectory must contain at least one state $\mathbf{s} \in \tilde{\mathcal{S}}_{g_1}$.

For this thesis, a novel approach has been developed which seamlessly and consistently integrates the combined multi-waypoint planning into the unified planning concept of hybrid dimensionality and multiple resolution planning. Early work used a tight integration with the graph search algorithm in order to jointly plan for multiple waypoints [Pet12]. However, the approach described in this thesis generalizes those ideas by augmenting the search space with an additional goal dimension [Pet13c]. For this

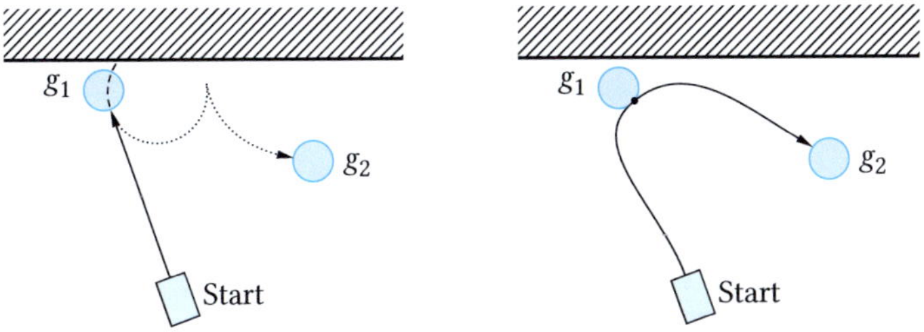

(a) Naive planning to next waypoint only.

(b) Combined planning for two waypoints.

Figure 3.15: Planning strategies for tasks with multiple waypoints.

purpose, the state (× time) lattice is extended with an additional dimension G, which is independent of the resolution and not affected by the dimensionality reduction process. Therefore, the lattice (3.14) is extended to

$$L^{d,r} = \left(\prod_{i=1}^{n} S_i^r \right) \times T^r \times G \,. \tag{3.72}$$

This results in

$$L^{0,r} = X^r \times Y^r \times \Theta^r \times V^r \times T^r \times G \,. \tag{3.73}$$

for the mobile robot motion planning example from Section 3.4. The set of goals

$$G = \{ g_1, \ldots, g_N \} \tag{3.74}$$

contains all waypoints of the current task (with N being the total number of waypoints of the task).

The goal-augmented lattice (3.72) consists of goal-augmented states

$$\tilde{\mathbf{s}}^{d,r} = (\tilde{s}_1, \tilde{s}_2, \ldots, \tilde{t}, g) \,. \tag{3.75}$$

The component g denotes the segment in which a state belongs; more precisely, g denotes the next waypoint during the planning. Thus, the search starts with a state

$$\tilde{\mathbf{s}}_{\mathrm{s}}^{d,r} = (\tilde{s}_{1,\mathrm{s}}, \tilde{s}_{2,\mathrm{s}}, \ldots, \tilde{t}_{\mathrm{s}}, g_1) \tag{3.76}$$

and ends with a goal state

$$\tilde{\mathbf{s}}_{\mathrm{g}}^{d,r} = (\tilde{s}_{1,\mathrm{g}}, \tilde{s}_{2,\mathrm{g}}, \ldots, \tilde{t}_{\mathrm{g}}, g_N)\,. \tag{3.77}$$

The transition between different goal segments during the graph search is explained in detail in Section 5.1.5.

3.9 Bringing It All Together

So far, this chapter has inductively described how a state lattice, being state of the art, can be extended with additional features in order to form a goal-augmented multi-resolution state × time lattice with hybrid dimensionality, which is the basis for efficient motion planning in unstructured dynamic environments. This section puts all the aspects together and explains the workflow for the application of the algorithm.

The first step is to decide on the relevant state variables for motion planning and to obtain the system model for the later generation of motion primitives. Secondly, the required planning fidelity has to be specified. This means one has to choose the number of dimensionality and resolution levels (i.e., the parameters $d_{\min}$ and $r_{\min}$) taking into consideration the trade-off between accuracy and planning speed. Furthermore, a decision has to be made on the discretization strategy for every state variable s_i. This includes choosing the respective quantization increments $\delta_{s_i}^r$ for each resolution level r.

Now, the motion primitive sets $\mathcal{M}^{0,0}, \ldots, \mathcal{M}^{d_{\min}, r_{\min}}$ can be sampled, which define the admissible motion of the dynamic system through the lattices $L^{0,0}, \ldots, L^{d_{\min}, r_{\min}}$ (see Section 3.5.4). The procedure is as follows. First, the high-dimensional motion primitive sets $\mathcal{M}^{0,0}, \ldots, \mathcal{M}^{0, r_{\min}}$ are generated. Care must be taken to ensure that motion primitive sets with fine resolution include motions from motion primitive sets with a more coarse resolution, i.e., condition (3.66) must be satisfied. Secondly, the dimensionality level is successively reduced (see Section 3.6). The first projection yields $\mathcal{M}^{1,0}, \ldots, \mathcal{M}^{1, r_{\min}}$. Once again, condition (3.66) needs to be ensured. This projection procedure is repeated until the lowest dimensionality level $d = d_{\min}$ is reached. With this, all required motion primitive sets have finally been obtained:

$$
\text{lower dim.}\Big\downarrow
\overset{\xrightarrow{\text{lower resolution}}}{
\begin{array}{ccc}
\mathcal{M}^{0,0} & \cdots & \mathcal{M}^{0,r_{\min}} \\
\vdots & \ddots & \vdots \\
\mathcal{M}^{d_{\min},0} & \cdots & \mathcal{M}^{d_{\min},r_{\min}}
\end{array}}
\tag{3.78}
$$

with their corresponding lattices

$$
\text{lower dim.}\Big\downarrow
\overset{\xrightarrow{\text{lower resolution}}}{
\begin{array}{ccc}
L^{0,0} & \cdots & L^{0,r_{\min}} \\
\vdots & \ddots & \vdots \\
L^{d_{\min},0} & \cdots & L^{d_{\min},r_{\min}}
\end{array}}
\tag{3.79}
$$

The elements of at least each first row, i.e., the motion primitive sets $\mathcal{M}^{0,0}, \ldots, \mathcal{M}^{0, r_{\min}}$ and the corresponding lattices $L^{0,0}, \ldots, L^{0, r_{\min}}$, contain the time among their dimensions; thus, they allow for explicit consideration of dynamic obstacles during the planning.

If multiple-waypoint planning is required, all lattices in (3.79) are augmented with an additional goal dimension (see Section 3.8). Finally, the search graph can be constructed on the basis of the generated motion primitives (3.78). This is explained in detail in Chapter 5.

3.10 Summary

In this chapter, a novel unified approach for multi-resolution robot motion planning with hybrid dimensionality has been presented [Pet13a; Pet13b; Pet13c; Pet14]. It is a consistent approach that transparently integrates time-parametrized motion planning as well as combined planning for multiple waypoints without the need for differentiation of special cases. The scientific contribution of this thesis is a novel rigorous formulation of a holistic method that enables the use of established and proven concepts like time-parametrized planning, multi-resolution planning, or planning with adaptive dimensionality together with new capabilities like multi-waypoint planning. The aim is not to target a specific application but to provide a universal approach for a wide variety of robotic motion planning tasks. The user has fine-grained control of the fidelity graduation desired for the actual usage scenario. For this purpose, a procedure for systematically obtaining state lattices with variable resolution and dimensionality was presented. This includes the automatic generation of the corresponding motion primitive sets from a given system model. For the sake of clarity, all these steps have been illustrated using a case study of motion planning for a mobile robot.

Chapter 4

Modeling the Environment

The approach developed in this thesis has been implemented on a robotic platform (see Section 6.4) based on the case study from Section 3.4. This chapter explains the modeling of the robot's environment so that this model can be used in Chapter 5 to construct the search graph on the basis of the lattice described in Chapter 3. Since the proposed planning scheme allows for time-parametrized planning, both the static environment and dynamic obstacles need to be modeled appropriately.

It is well known that most of the computation time during motion planning is spent on collision checking [LaV06]. Thus, efficient collision checking is an active research field of its own and hence is not the focus of this thesis. Nonetheless, there need to be means available for fast collision checking for real-world application of the proposed motion planning algorithm. Therefore, this chapter only touches the topic and briefly describes the collision checking strategies used in this thesis in order to facilitate the sound interpretation of the results in Chapter 6.

The proposed method distinguishes three reasons for collisions, namely static obstacles, terrain characteristics, and dynamic obstacles. It is an attempt to establish a unified, consistent model that incorporates the perception and prediction uncertainty in a probabilistic formulation.

4.1 Static Obstacles

There are many ways to acquire a model of the robot's surroundings. In an unknown outdoor environment—which, for example, needs to be explored after a natural disaster has struck—a preliminary drivability map may be obtained from aerial imagery by photogrammetric methods [Cur15]. The actual execution of the motion is generally based on an up-to-date local map $\mathcal{W}$ of the environment, which is obtained by a LIDAR[1] scanner. The map is commonly encoded in terms of a grid consisting of both free and occupied cells. It is the collision checker's task to determine whether the robot $\mathcal{A}$ with its configuration $\mathbf{q}$ is in a collision with any occupied cell of the grid map (cf. (1.1)). For this purpose, many methods have been proposed, which most notably differ in the way they handle the robot's geometry.

One approach for collision checking is the convolution of the whole grid map with the robot's footprint for each possible orientation, which results in multiple map slices [Zie08; Lin08]. Then, to check for a collision during the planning, one must only pick the appropriate map slice and perform a simple look-up of the cell that corresponds to the robot's position. This method, however, is disadvantageous for large maps because it is always the whole map that needs to be convolved with the robot's footprint several times. For this reason, this method is particularly well suited for stationary environments that do not change between planning cycles.

This thesis focuses on applications for unknown, unstructured environments, and it is therefore assumed that the entire map may change

1. Light detection and ranging.

between planning cycles during exploration as more and more information becomes available. This condition can be regarded as a worst-case scenario. Thus, in the context of this thesis, *static environment* refers to the part of the environment that does not move as opposed to the dynamic obstacles that do move such as people or other robots.

Another approach for collision checking is presented in [Fer08b; Lik09]: The authors precompute convolution kernels for all robot orientations and then perform on-demand collision checking during the planning for each robot pose by convolution of the appropriately shifted kernel with the map. In order to avoid unnecessary computational work, the following two checks are performed in advance: If there is a collision with the inscribed circle of the robot footprint, the actual robot footprint is also in a colliding configuration irrespectively of the robot's orientation. If even the circumscribed circle of the robot's footprint is collision-free, then the actual robot footprint is also guaranteed to be collision-free. In both cases, computationally expensive convolution with the exact robot footprint can be avoided. A further approach to speed up the costly convolution process is proposed by [Zie10]. The authors approximate the robot footprint by multiple disks and then perform a fast convolution by exploiting an axis-aligned rectangle representation of the disks. A similar method, which has been proposed in [Kin09], uses a decomposition of the robot's footprint into a set of disks and a set of remaining cells which are not covered by the disks.

Collision checks are commonly performed for discrete points in time along the examined trajectory. It is, however, also possible to assess the entire region swept over by the robot (called *swath*) when executing a motion primitive [Pet12; Piv09c].

4.1.1 Distance-Based Collision Checking

The demonstrator that has been developed during this thesis has an almost square footprint; hence, it is legitimate to approximate its shape with a disk and perform orientation-independent collision checking. This is done on the basis of the exact Euclidean distance transform (EDT) of the grid map, which is computed at the beginning of each planning cycle by the algorithm proposed in [Mau03]. The EDT assigns each cell of the map the Euclidean distance to the next occupied cell (see Figure 4.1 for an example). With this, a collision can be easily checked for by inspecting the cell corresponding to the robot's position: If the distance to the next obstacle is less than the radius of the robot's circumscribing circle, the robot is assumed to be in a colliding state. This distance may be enlarged by an additional safety zone (see Section 4.1.2) if desired. The trajectories (3.39) of the respective motion primitives (3.41) have been incrementally sampled with the temporal resolution δ_t^r in Section 3.5.4. This allows an efficient collision checking for each intermediate state of the trajectory $\mathcal{T}$ of a motion primitive m for each increment δ_t^r. Of course, it has to be ensured that δ_t^r is small enough in order not to skip over obstacles during collision checking.

4.1.2 Risk Zones

In addition to simple collision checking, it can be useful to consider a generalized collision risk with the static environment. This allows for an intuitive modeling of the fact that collisions are more likely the closer the robot gets to an obstacle. This can be used to create a safety zone around obstacles to account for uncertainties in perception, localization, and in the execution of the planned motion. The safety zone is defined on the basis of the result of the exact Euclidean distance transform in the previous section. It assigns each cell of the map a probability p_{stat} for collision with

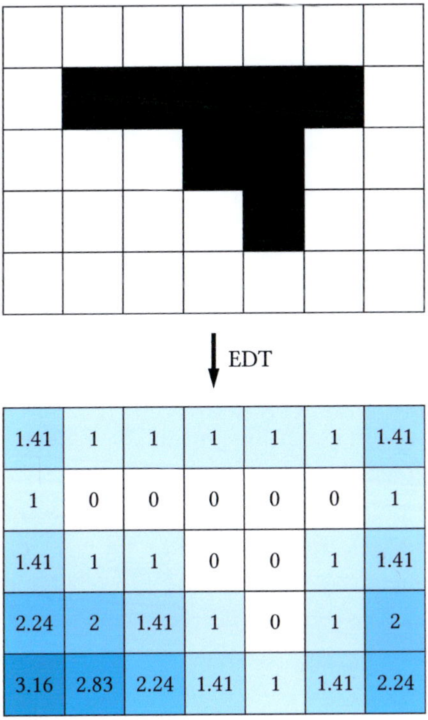

Figure 4.1: Exact Euclidean distance transform (EDT). Top: occupancy grid map (■ occupied, □ free). Bottom: distance map after application of EDT.

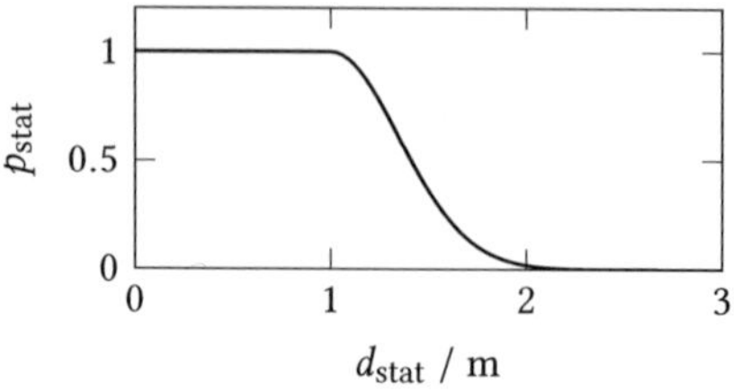

Figure 4.2: Collision probability (4.1) with static obstacles for $\rho = 1\,\text{m}$ and $\gamma = 4\,\text{m}^{-2}$.

a static obstacle on the basis of the distance d_{stat} to the next occupied cell:

$$p_{\text{stat}}(d_{\text{stat}}) := \begin{cases} 1 & \text{if } d_{\text{stat}} \leq \rho \\ e^{-\gamma(d_{\text{stat}}-\rho)^2} & \text{else} \end{cases} . \tag{4.1}$$

The parameter ρ denotes the minimum distance to the next obstacle. This minimum distance needs to be observed at all times, i.e., a distance below this threshold results in a collision probability of 100 %. The parameter γ denotes the rate at which the collision probability decreases with increasing distance from the obstacle. By evaluating p_{stat} according to (4.1), the separate collision check from Section 4.1.1 becomes redundant and can thus be omitted. Figure 4.2 shows an example of a typical collision probability for $\rho = 1\,\text{m}$. The complete process of computing the risk zones is visualized in Figure 4.3.

4.2 Terrain

Especially for planning in rough, unstructured outdoor environments, it is important to take the characteristics of the terrain into account, since there may be regions that are not particularly well suited for driving. In order to identify such regions, methods that determine the roughness of

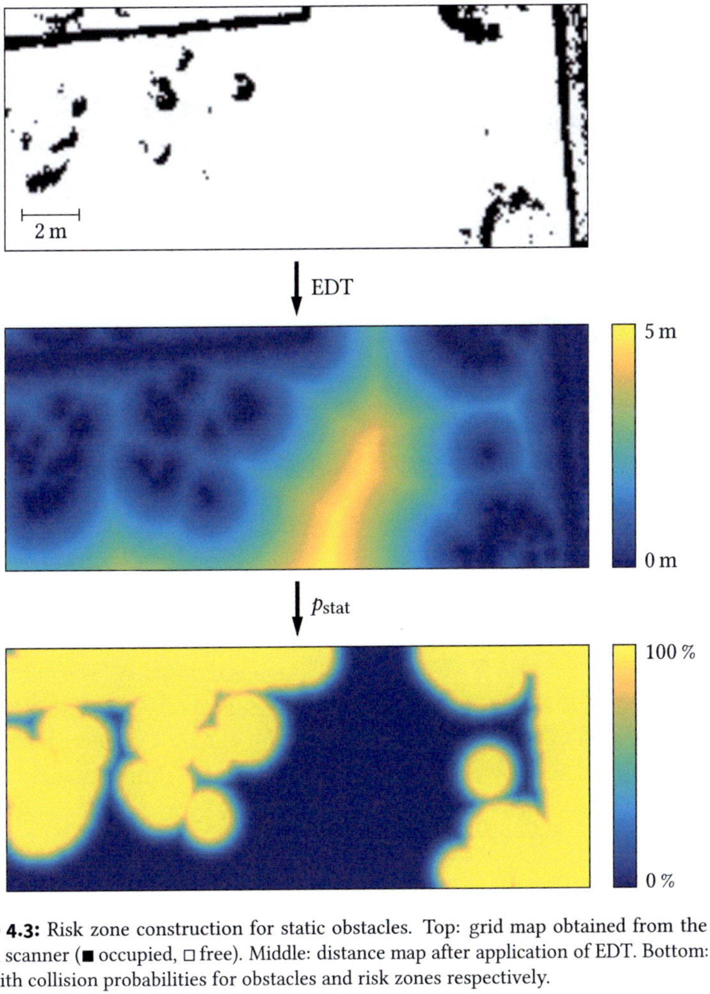

Figure 4.3: Risk zone construction for static obstacles. Top: grid map obtained from the LIDAR scanner (■ occupied, □ free). Middle: distance map after application of EDT. Bottom: map with collision probabilities for obstacles and risk zones respectively.

the ground using LIDAR data can be employed [Neu09]. The classification of the terrain allows for each cell—similar to the previously described risk zones—an association with a probability p_{terrain} for collision due to rough terrain.

In addition, it is, at this point, possible to integrate a priori knowledge of the environment into the model by manually defining static hazardous regions with increased collision risk [Pet13c]. This may be, for example, interesting in a logistics application, where forklift trucks cause heavy traffic on a route known in advance. In such a scenario, the robot should avoid such regions where feasible, or, if the robot's task leads through such a region, the robot should cross it using the shortest way possible (see Figure 6.17). These custom regions with increased collision probability are handled in the same way the terrain characteristics are handled and are thus also modeled by the respective collision probability p_{terrain}.

4.3 Dynamic Obstacles

Large portions of this thesis have been developed in the context of the SENEKA project [Kun14]. The main goal of SENEKA is the fast and effective reconnaissance of disaster areas by human-robot teams. Besides the potentially rough terrain, the robot needs to be able to cope with moving participants in the scene. These may be, for example, members of rescue teams or other vehicles, both human-controlled or autonomously operating. All these objects are collectively referred to as *dynamic obstacles* in the following. They are assumed to move in a—from a planning perspective—non-cooperative way. This is in contrast to scenarios like cooperative planning for road vehicles [Fre11].

A lot of research has been conducted in the past on avoiding collisions with dynamic obstacles. For example, the authors of [Fra04b] propose the concept of *inevitable collision states*, which are defined as regions of the state space that will inevitably lead to a collision in the future. The inevitable collision states are computed on the basis of set-theoretic considerations for obstacles of arbitrary shape. Unfortunately, the authors do not readily supply any information on the computational expense of their method. In order to perform collision checking in real-time, a common approach is to approximate the dynamic obstacles' footprint by a disk. For example, in [Ber06], the disk-shaped footprint of the obstacles is propagated in time on the basis of their estimated trajectory and, for an efficient collision checking, the robot's footprint is assumed to be circular, too. Another approach is the integration of fake obstacles into the static obstacle map by predicting the dynamic obstacle's motion for a limited duration and determining the swept over area [Fer08b]. In this way, it is, of course, impossible to perform truly time-parametrized motion planning among dynamic obstacles. Furthermore, many collision checking approaches simply ignore the fact that the present—not to mention future—position of a dynamic obstacle is not precisely known because it is only estimated by some form of obstacle tracker. Therefore, this thesis focuses on a probabilistic modeling of dynamic obstacles in order to integrate them into a consistently formulated concept of collision probability computation.

4.3.1 Probabilistic Modeling of Disk-Shaped Obstacles

One collision checking method that was implemented in this thesis assumes approximately circular dynamic obstacles. This assumption is, for example, justified if people make up most of the dynamic objects in a scene. The collision checking is based on the method presented in [Kus09] but adds some extensions for better performance and accuracy. The method relies on the availability of a perception module that provides an estimation of the obstacles' states together with the corresponding uncertainty. The state variables are assumed to be normally distributed random variables. With an appropriate system model (e.g., a simple constant velocity model), this state can be used to predict the future trajectory of the dynamic obstacle and the evolution of the associated uncertainty. For this purpose, the standard prediction step of an extended Kalman filter [Kle12] may be used if the obstacle's linear (or even rotational) velocity is known from the obstacle tracker.

For this kind of modeling, the risk for colliding with a dynamic obstacle depends on the position, orientation, size, and uncertainty (covariance matrix with six distinct elements) of the obstacle as well as on the position and orientation of the robot. This results in a 13-dimensional parameter vector, which prevents any reasonable precomputation of collision risks. Because of this, dynamic obstacles are reduced to point obstacles and the original extent of a dynamic obstacle is added to the size of the circular robot footprint. With this simplification, the number of parameters necessary to define the probability for a collision with the i-th obstacle is reduced to six, namely the relative position of robot and obstacle, $d_{\mathrm{dyn},i} = [d_{\mathrm{dyn},x,i}\ d_{\mathrm{dyn},y,i}]^\top$, the predicted covariance matrix

$$\Sigma_i = \begin{bmatrix} \sigma_{xx,i} & \sigma_{xy,i} \\ \sigma_{xy,i} & \sigma_{yy,i} \end{bmatrix}, \tag{4.2}$$

consisting of three distinct values, and the enlarged robot footprint radius ζ_i, which is the sum of the robot radius and the obstacle radius. The computation of the probability for colliding with the i-th obstacle at a particular instant of time is done by integrating the predicted probability distribution of the dynamic obstacle over the enlarged robot footprint R_i:

$$p_{\mathrm{dyn},i} = \iint\limits_{R_i} \frac{1}{2\pi\sqrt{|\Sigma_i|}} \exp\left(-\frac{1}{2}\begin{bmatrix} x & y \end{bmatrix} \Sigma_i^{-1} \begin{bmatrix} x \\ y \end{bmatrix}\right) dx\, dy \tag{4.3}$$

with

$$R_i = \left\{ (x, y) \,\middle|\, (x - d_{\mathrm{dyn},x,i})^2 + (y - d_{\mathrm{dyn},y,i})^2 \le {\zeta_i}^2 \right\}. \tag{4.4}$$

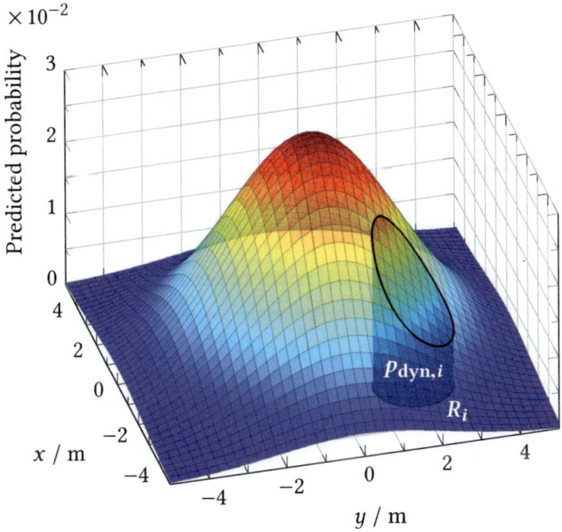

Figure 4.4: Calculation of collision probability (4.3) [Pet13a].

The probability (4.3) for collision with a single dynamic obstacle needs to be computed for each dynamic obstacle individually. Figure 4.4 illustrates this computation. The combined probability for colliding with any of N present obstacles is obtained by the joint probability

$$p_{\text{dyn}} = 1 - \prod_{i=1}^{N} (1 - p_{\text{dyn},\,i}) \,, \tag{4.5}$$

which exploits the simplifying assumption that each collision is probabilistically independent.

The authors of [Kus09] use a look-up table with precomputed values of (4.3) for common parameter values. However, a six-dimensional look-up table allows only a limited resolution for the quantization of each parameter. For example, already 25 distinct values for each parameter would lead to a total of 2.4×10^8 entries in the look-up table requiring 1 GB of memory when stored as 4-byte floating-point numbers. For this reason, an approach that computes the integral (4.3) on demand and caches the result for each triple (x, y, t) visited during a planning cycle has been developed in this thesis [Pet13c]. The proposed approach does not restrict the collision probability computation to a limited set of robot-obstacle constellations thus increasing the accuracy. The caching is a good compromise between exhaustive precomputation and permanent recomputation of (4.3). Figure 6.15 shows an example for motion planning in the presence of a circular dynamic obstacle.

4.3.2 Probabilistic Modeling of Obstacles with Arbitrary Shape

The second collision checking method that was developed in this thesis can handle dynamic obstacles of arbitrary shape. This is especially useful for scenarios containing dynamic obstacles like cars or trucks with footprints substantially deviating from a circular shape. Furthermore, the proposed method does not impose any restrictions on the robot's shape. The approach aims at precise modeling of each object and its collision probability with fast and efficient means for computation in mind.

The method assumes a polygonal description of both the robot and dynamic obstacles. It is not restricted to convex polygons; however, most robots and obstacles feature a convex shape. The polygons serve mainly as the input for the algorithm; most of the collision risk computation is performed using a grid representation. First, some parameters need to be determined that control the granularity of the collision risk computation: the resolution Δ_{xy} of the grids that are used in the following computation process, the number of distinct robot orientations for collision checking, and the temporal granularity Δ_t for the prediction of the dynamic obstacles.

The grid resolution Δ_{xy} can be chosen independently of the map resolution or the state lattice quantization δ^r_{xy}, and also the number of robot orientations is not restricted to the previously chosen set of admissible robot headings Θ, see (3.27). On the other hand, it makes sense to choose the temporal granularity Δ_t in accordance with the temporal increment δ_t, see (3.13), which has also been used in the motion primitive sampling process (see Section 3.5.4).

Robot masks

Once during the initialization of the algorithm, a set of robot masks needs to be created. For any of the considered orientations the robot polygon is rotated and then converted to a grid representation (i.e., a matrix)

$$\mathrm{R}(\theta) = (r_{ij}), \quad \mathrm{R} \in \mathbb{R}^{m(\theta)\times n(\theta)} \tag{4.6}$$

with occupied cells set to 1 and free cells set to 0. This rasterization process can be efficiently performed by using a scan-line algorithm, which is a well-known image processing technique [Hug14]. The robot masks are stored in memory and will serve as a stencil in order to compute the region simultaneously occupied by the robot and an obstacle in the following collision risk computation.

Time slices

For each planning cycle, a set of *time slices* is generated containing information on the location of the dynamic obstacles. A time slice is a square matrix

$$\Omega(t) = (\omega_{ij}), \quad \Omega \in \mathbb{R}^{M(t)\times M(t)}, \quad M(t)\ \text{odd}, \tag{4.7}$$

which is a grid representation of the xy-plane at a particular time $t = n'\Delta_t$, $n' \in \mathbb{N}^+$. The total number of time slices depends on the temporal horizon for planning with full dimensionality in the state × time lattice (see Section 5.1.2).

Each time slice needs to be large enough to cover the robot's entire radius of action at time t. Thus, the size $M(t)$ of time slice $\Omega(t)$ is determined by

$$M(t) := 2\lceil d_{\max}(t)/\Delta_{xy}\rceil + 1 \tag{4.8}$$

with $d_{\max}(t) = v_{\max}t$, where $v_{\max}$ is the maximum velocity of the robot (see Figure 4.5). Consequently, the size of the time slices increases with time. Each time slice is centered on the robot's position, (x_0, y_0), at the start of the planning cycle. This provides a global referencing with the environment and its objects. If there are multiple dynamic objects in a scene, a separate tentative stack of time slices is generated for each obstacle.

The polygon of each dynamic obstacle is predicted forward in time on the basis of the information provided by the perception module. Depending on the available motion hypotheses, various transformations are possible, like translations and rotations. Theoretically, even the shape of the dynamic obstacles may change over time; however, this is rarely encountered in practice. After the transformation for a particular time t, the new polygon is rasterized into the corresponding time slice. The value of each cell of a time slice can be interpreted as the probability of being part of an obstacle, which means that occupied cells are assigned the value 1 and free cells are assigned the value 0.

The estimation and prediction of the obstacles' trajectories is generally associated with a certain degree of uncertainty. This information is utilized by performing a convolution of each time slice with a Gaussian smoothing kernel, whose variance is determined on the basis of the estimated uncertainty. As an example, Figure 4.5 shows the resulting stack of time slices for one obstacle after this process.

Finally, if there are N obstacles, their respective stacks of time slices $\Omega_k(t)$, $k = 1, \ldots, N$ are merged into one single stack by combining con-

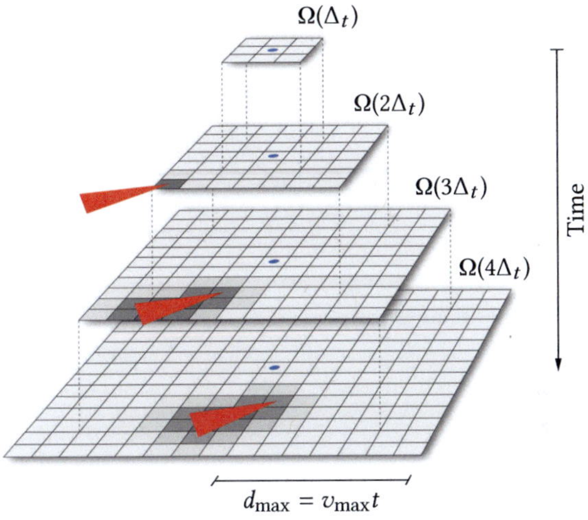

Figure 4.5: Stack of time slices for one obstacle (red triangle). The slices are centered around the current robot position (blue dot).

temporaneous time slices into one:

$$\Omega(t) = \left(\omega_{ij}(t)\right) = \left(1 - \prod_{k=1}^{N}\left(1 - \omega_{k,ij}(t)\right)\right). \tag{4.9}$$

This is equivalent to (4.5) and assumes that the events of being occupied by the k-th obstacle are probabilistically independent. This concludes the preparatory computations at the beginning of each planning cycle.

Collision probability

The actual collision probability computation is performed for each sampled state of the motion primitives' trajectories (3.39). First, the time slice that corresponds to the time associated with the currently examined state needs to be selected. Next, the robot mask is translated according to the current robot position and then superimposed on the time slice. Finally, the sum of the values of all cells that are covered by the robot is computed:

$$p_{\text{dyn}}(x, y, \theta, t) = \min\left(1, \sum_{i=1}^{m(\theta)} \sum_{j=1}^{n(\theta)} r_{ij}(\theta)\, \omega_{i+i'(x,t), j+j'(y,t)}(t)\right), \tag{4.10}$$

with m and n denoting the size of the robot mask and

$$i'(x, t) = \left\lfloor \frac{x - x_0}{\Delta_{xy}} \right\rceil + \frac{M(t) - 1}{2} - \frac{m(\theta) - 1}{2}, \tag{4.11}$$

$$j'(y, t) = \left\lfloor \frac{y - y_0}{\Delta_{xy}} \right\rceil + \frac{M(t) - 1}{2} - \frac{n(\theta) - 1}{2}, \tag{4.12}$$

where (x_0, y_0) is the robot's position at the start of the planning cycle and $\lfloor \cdot \rceil$ denotes rounding to the nearest integer. The computation corresponds to the correlation of robot mask and time slice evaluated at the current position of the robot. For point obstacles (i.e., obstacles without physical extent) the evaluation of (4.10) yields the collision probability $p_{\text{dyn}} = 1$ if all time slice cells that are affected by the dynamic obstacle are located inside the robot footprint. For spatially extended obstacles, the sum in (4.10) may become larger than 1. This would prevent the utilization of (4.10) as a probability measure, which is why the sum is clipped to 1. This approximate approach for a probabilistic interpretation of the collision probability allows a quick assessment of the collision risk and proved its worth in practice (see Section 6.3.3).

4.4 Overall Collision Risk

The overall collision probability is evaluated for each state of the trajectory $\mathcal{T}$ of each motion primitive m expanded during the planning (see Section 5.3). It is assumed that the events of colliding due to static obstacles, terrain characteristics, or dynamic obstacles are mutually independent. Thus, the overall collision probability p_{coll}, i.e., the collision risk, can be obtained by (cf. [Pet13c])

$$p_{\text{coll}} = 1 - (1 - p_{\text{stat}})(1 - p_{\text{terrain}})(1 - p_{\text{dyn}}) \,. \tag{4.13}$$

It is composed of the probability p_{stat} for colliding with static obstacles according to (4.1), the probability p_{terrain} for colliding due to terrain characteristics according to Section 4.2, and the probability p_{dyn} for colliding with dynamic obstacles according to (4.5) or (4.10).

4.5 Summary

This chapter gave a short overview of the methods that have been developed during this thesis for modeling the environment of a mobile robot. This has been necessary in order to put the proposed planning approach into practice and build a real-world demonstrator (see Section 6.4) that is capable of autonomous driving in unstructured and dynamic environments. For this purpose, a probabilistic approach based on the computation of a collision risk has been developed. This collision risk consists of three components, namely the collision probability with static obstacles, the collision probability due to terrain characteristics, and the collision probability with dynamic obstacles. For the latter, two alternative approaches have been developed: an extended version of the method that has been proposed in [Kus09] for circular obstacles and a novel method for obstacles with arbitrary shape, which is based on a grid representation in the form of time slices. This environment model provides the basis for the cost computation during the graph search (see Section 5.3).

Chapter 5

Searching the Lattice

Chapter 3 established the building blocks for mobile robot motion planning in unstructured dynamic environments, namely the concept of a multi-resolution state × time lattice with hybrid dimensionality and the corresponding sets of motion primitives that define the admissible motion conforming to the lattice structure. This chapter explains how the building blocks can be used to construct a search graph for finding the optimal motion plan. The search for the optimal path/trajectory of the robot is converted to the problem of finding the shortest path in a graph. The underlying lattice structure guarantees that the planning is complete (i.e., if a solution exists, it will be found, and if there is no solution, the algorithm will report this fact in finite time). Furthermore, the graph search algorithm guarantees that the motion planning is optimal (with respect to the employed resolution and the sampled motion primitives). A lot of well-known algorithms exist for finding shortest paths in a graph. However, especially in the presence of dynamic obstacles, quick solutions to the planning problem are vital in order to guarantee collision-free motion of the robot. This imposes significant demands on the graph search algorithm.

5.1 Building the Search Graph

The search graph is constructed by the concatenation of motion primitives. In order to define admissible combinations of motion primitives, the notation of *compatibility* is borrowed from [Fra05], where it is used in the context of an automaton-based maneuver planning algorithm. The notion is adapted to fit the method proposed in this thesis and is extended for the advanced concepts of variable resolution and dimensionality.

5.1.1 Motion Primitive Concatenation without Changing Dimensionality or Resolution

To illustrate the concept of *compatibility*, concatenation of motion primitives without changing the resolution level r or dimensionality level d is considered first. A discrete state $\tilde{\mathbf{s}}^{d,r} \in L^{d,r}$ is said to be compatible with a motion primitive

$$m^{d,r} = \left(\tilde{\mathbf{s}}_0^{d,r}, \tilde{\mathbf{s}}_\mathrm{e}^{d,r}, \mathbf{u}, \mathcal{T}, \Delta t_\mathrm{m}\right) \in \mathcal{M}^{d,r} \tag{5.1}$$

if there exists a translation vector $\delta\mathbf{s}$ such that

$$\tilde{\mathbf{s}}^{d,r} = \tilde{\mathbf{s}}_0^{d,r} + \delta\mathbf{s}\,, \tag{5.2}$$

where $\delta\mathbf{s}$ may only have nonzero components δs_i for the translation-invariant state variables s_i (see Section 3.5.2 for the definition of translational invariance). Following the notation of [Fra05], this compatibility relation is briefly written as

$$\tilde{\mathbf{s}}^{d,r} C\, m^{d,r}\,. \tag{5.3}$$

The application of a compatible motion primitive $m^{d,r}$ to a state $\tilde{\mathbf{s}}^{d,r}$, denoted by $\tilde{\mathbf{s}}^{d,r} m^{d,r}$, can be thought of as a mapping $L^{d,r} \times \mathcal{M}^{d,r} \to L^{d,r}$.

The motion primitive $m^{d,r}$ is first translated by the vector $\delta\mathbf{s}$ according to the compatibility condition (5.2), which results in a new motion primitive

$$m^{d,r\prime} = \left(\tilde{\mathbf{s}}_0^{d,r} + \delta\mathbf{s},\ \tilde{\mathbf{s}}_\mathrm{e}^{d,r} + \delta\mathbf{s},\ \mathbf{u},\ \mathcal{T}_{\delta\mathbf{s}},\ \Delta t_\mathrm{m}\right). \tag{5.4}$$

Then, the new discrete state $\tilde{\mathbf{s}}^{d,r\prime}$ after the application of $m^{d,r}$ is given by the new end state

$$\tilde{\mathbf{s}}^{d,r\prime} = \tilde{\mathbf{s}}^{d,r} m^{d,r} = \tilde{\mathbf{s}}_\mathrm{e}^{d,r} + \delta\mathbf{s}\,. \tag{5.5}$$

The transformed trajectory $\mathcal{T}_{\delta\mathbf{s}}$ is obtained by shifting each state $\mathbf{s} \in \mathcal{T}$ individually by the translation vector $\delta\mathbf{s}$ and provides the basis for the collision checking.

When a state $\tilde{\mathbf{s}}^{d,r}$ is expanded during the graph search, its successors are obtained by application of all compatible motion primitives, i.e., the set of successor states is given by

$$\mathrm{succ}\left(\tilde{\mathbf{s}}^{d,r}\right) = \left\{ \tilde{\mathbf{s}}^{d,r} m^{d,r} \;\middle|\; \tilde{\mathbf{s}}^{d,r} C\, m^{d,r} \wedge m^{d,r} \in \mathcal{M}^{d,r} \right\}. \tag{5.6}$$

5.1.2 Transitions Between Different Dimensionalities

Gradually reducing the planning dimensionality is a key idea of the approach that has been developed in this thesis (see Section 3.6). This allows high-fidelity planning for the immediate future and more coarse planning for more distant regions. The desired dimensionality is selected on the basis of the elapsed time to reach the considered state [Pet14]. For this purpose, the time $\bar{t}(\tilde{\mathbf{s}}^{d,r})$ for reaching a state $\tilde{\mathbf{s}}^{d,r}$ in the search graph is stored alongside the state $\tilde{\mathbf{s}}^{d,r}$. The time $\bar{t}(\tilde{\mathbf{s}}^{d,r})$ is the sum of the durations of the individual motion primitives leading to this state. The application of the motion primitive $m^{d,r} = (\tilde{\mathbf{s}}_0^{d,r}, \tilde{\mathbf{s}}_\mathrm{e}^{d,r}, \mathbf{u}, \mathcal{T}, \Delta t_\mathrm{m})$ to a state $\tilde{\mathbf{s}}^{d,r}$ leads to a new state $\tilde{\mathbf{s}}^{d,r\prime}$ with

$$\bar{t}(\tilde{\mathbf{s}}^{d,r\prime}) = \bar{t}(\tilde{\mathbf{s}}^{d,r}) + \Delta t_\mathrm{m}\,. \tag{5.7}$$

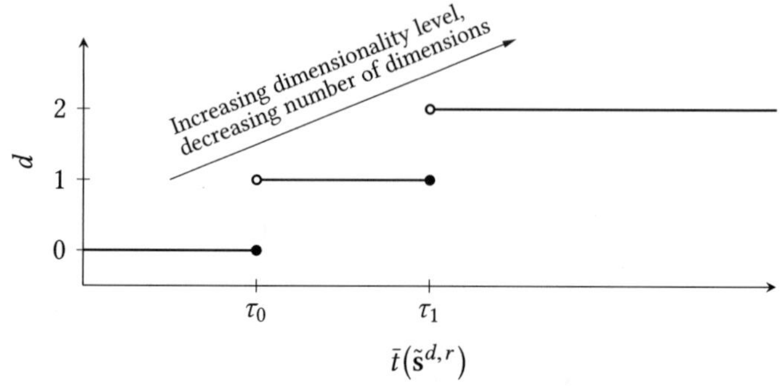

Figure 5.1: Temporal thresholds for dimensionality reduction.

During the planning, the employed dimensionality level is controlled by temporal thresholds τ_d with $d \in \{0, \ldots, d_{\text{min}} - 1\}$ and $\tau_d \leq \tau_{d'}$ for $d < d'$. Starting with $d = 0$, a dimensionality level d is used until the corresponding temporal threshold τ_d, at which the transition to a lower dimensionality occurs. This is illustrated in Figure 5.1.

The dimensionality transition in the search graph is performed during the appropriate generation of a state's successors. If a dimensionality reduction is pending for the currently expanded state $\tilde{\mathbf{s}}^{d,r}$ and time $\bar{t}(\tilde{\mathbf{s}}^{d,r})$, i.e., if $\bar{t}(\tilde{\mathbf{s}}^{d,r})$ exceeds τ_d, the successors of $\tilde{\mathbf{s}}^{d,r}$ are determined not on the basis of this very state $\tilde{\mathbf{s}}^{d,r}$ but on the basis of the projected state $\tilde{\mathbf{s}}^{d+1,r} = \pi(\tilde{\mathbf{s}}^{d,r})$. Thus, when switching the dimensionality level, the successors of a state $\tilde{\mathbf{s}}^{d,r}$ are given, by analogy to (5.6), as

$$\operatorname{succ}\left(\tilde{\mathbf{s}}^{d,r}\right) = \left\{ \pi\left(\tilde{\mathbf{s}}^{d,r}\right) m^{d+1,r} \,\middle|\, \pi\left(\tilde{\mathbf{s}}^{d,r}\right) C\, m^{d+1,r} \wedge m^{d+1,r} \in \mathcal{M}^{d+1,r} \right\}, \tag{5.8}$$

where $m^{d+1,r}$ is now an element of the motion primitive set $\mathcal{M}^{d+1,r}$ with reduced dimensionality.

In the context of the mobile robot example from Section 3.4, the dimensionality reduction essentially means the following: From $\bar{t} = 0$ to $\bar{t} = \tau_0$ a full-dimensional planning including the temporal dimension is performed. This allows for true time-parametrized motion planning among dynamic obstacles. At $\bar{t} = \tau_0$ the temporal dimension is dropped and the planning proceeds with dynamically feasible planning (i.e., still considering the robot's dynamics) until $\bar{t} = \tau_1$. Finally, at $\bar{t} = \tau_1$ the velocity dimension is also dropped, and the planning changes to only kinematically feasible planning until reaching the goal.

5.1.3 Transitions Between Different Resolutions

The employed planning resolution is the second possibility for adjusting the planning fidelity. The criteria for a reasonable resolution selection are described in Section 5.1.4. This section explains the change of the resolution during the search graph construction [Pet13b]. The currently used resolution is specified by the resolution level r (see Section 3.7). In contrast to the dimensionality reduction, a change of resolution may be made in any direction, i.e., both to more coarse as well as to finer resolutions.

With the compatibility relation (5.2), the transition between different resolutions can be concisely formulated: If a transition from a state $\tilde{\mathbf{s}}^{d,r}$ with corresponding resolution level r to the new resolution level r' is intended, the successors of $\tilde{\mathbf{s}}^{d,r}$ can be obtained by

$$\mathrm{succ}\left(\tilde{\mathbf{s}}^{d,r}\right) = \left\{ \tilde{\mathbf{s}}^{d,r} m^{d,r'} \,\middle|\, \tilde{\mathbf{s}}^{d,r} C\, m^{d,r'} \wedge m^{d,r'} \in \mathcal{M}^{d,r'} \right\} . \tag{5.9}$$

If $r' < r$, i.e., a transition is made to a region with finer resolution, the prerequisites (3.66) and (3.67) guarantee that $\tilde{\mathbf{s}}^{d,r} \in L^{d,r}$ is also an element of $L^{d,r'}$ and thus motion primitives compatible with $\tilde{\mathbf{s}}^{d,r}$ exist in $\mathcal{M}^{d,r'}$. On the other hand, if the resolution is decreased, i.e., $r' > r$, the state $\tilde{\mathbf{s}}^{d,r}$ is not necessarily also an element of $L^{d,r'}$, which may result in an empty set of successors (5.9). In this case, the currently expanded branch of the graph is closed. Thus, a transition from $\tilde{\mathbf{s}}^{d,r} \in L^{d,r}$ to a state $\tilde{\mathbf{s}}^{d,r'} \in L^{d,r'}$ with a more coarse resolution can only happen if $\tilde{\mathbf{s}}^{d,r}$ is already an element of $L^{d,r'}$.

The change of resolution is totally independent of the dimensionality reduction process. Moreover, both may occur at the same time, and the successors of $\tilde{\mathbf{s}}^{d,r}$ are then obtained by the combination of (5.8) and (5.9),

$$\operatorname{succ}\left(\tilde{\mathbf{s}}^{d,r}\right) = \left\{ \pi\left(\tilde{\mathbf{s}}^{d,r}\right) m^{d+1,r'} \;\middle|\; \pi\left(\tilde{\mathbf{s}}^{d,r}\right) C\, m^{d+1,r'} \wedge m^{d+1,r'} \in \mathcal{M}^{d+1,r'} \right\}. \tag{5.10}$$

5.1.4 Regions of High-Resolution Planning

A particular planning resolution is associated with each part of the search space. The planning scheme proposed in this thesis allows an arbitrary number of resolution levels; however, in practice, two resolution levels are often sufficient. Thus, this section describes the methods that are used to determine the low- and high-resolution planning regions of the search space. The selection of the desired resolution can be realized on the basis of various criteria. The authors of [Ruf09a] suggest the differentiation between *task-based* and *environment-based* criteria.

Task-based criteria are the most commonly employed criteria, and have been, for example, also used by *Boss* in the DARPA Urban Challenge [Lik09]. They basically define high-resolution planning regions for the vicinity of the robot and the goal. This allows high-fidelity planning for the immediate future as well as for intricate goal configurations, e.g., in a parking maneuver. Due to the initial high-resolution planning region, the continuous replanning scheme produces trajectories with an overall quality similar to what would have been achieved by planning in a high-resolution lattice only [Ruf09a].

Environment-based criteria form the second class of criteria. Many different strategies for defining high-resolution regions are possible. For example, one may opt for high-resolution planning near obstacles or in areas with increased collision risk. These regions can be easily obtained from the Euclidean distance transform (EDT) of the occupancy map (see Section 4.1) and the corresponding collision probability p_{stat}. In order to reduce the extent of high-resolution planning regions and thus the computation time, it might be reasonable to restrict the high-resolution areas to narrow passages. In this way, it can be ensured that solutions leading through narrow passages are not missed because of too coarse planning. The authors of [Ruf09a] use the so-called *bridge test* [Sun05] for detection of narrow passages. This randomized algorithm has been developed in the context of Probabilistic Roadmaps (PRMs) and is especially useful for high-dimensional configuration spaces. The obstacle maps for mobile robot motion planning applications, however, are usually two-dimensional occupancy grid maps and therefore a different, morphology-based approach has been implemented in this thesis in order to enable fast and deterministic identification of narrow passages. For this purpose, the occupancy grid map is first dilated with an appropriately sized structuring element (depending on the size of the robot) and subsequently eroded with the same structuring element. This corresponds to the well-known morphological *closing* operation [Har87]. The difference of the resulting grid map and the original grid map marks the narrow passages and hence high-resolution planning regions (see Figure 5.2). The map of the desired planning resolution is stored alongside the obstacle map and is evaluated for each expanded state during the graph search.

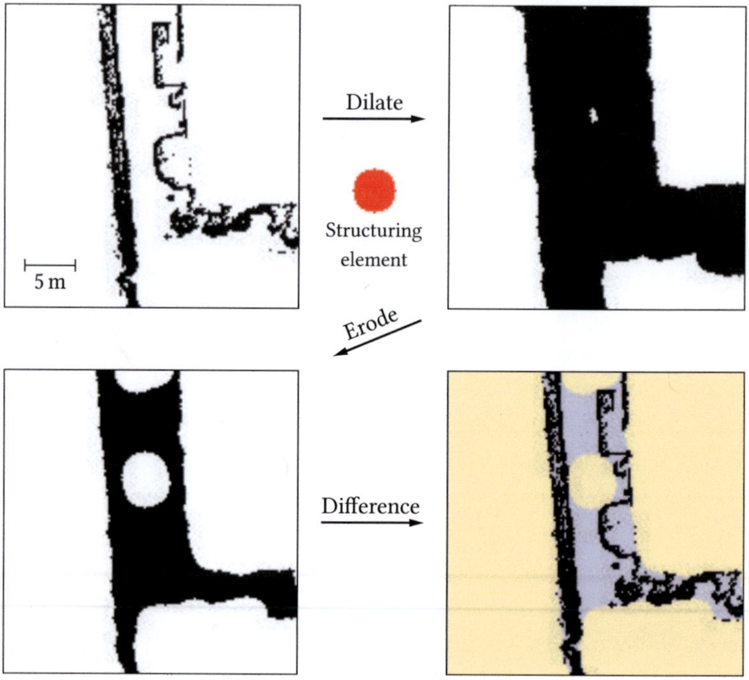

Figure 5.2: Determination of high-resolution planning regions. Top left: the original occupancy grid map. Top right: dilated occupancy grid map. Bottom left: occupancy grid map after *closing* operation. Bottom right: the high-resolution planning region is obtained by the difference of closed and original occupancy map (■ obstacle, □ low-resolution planning region, □ high-resolution planning region).

5.1.5 Multiple Waypoints

Planning for multiple waypoints was introduced in Section 3.8. For this purpose, each state $\tilde{\mathbf{s}}^{d,r}$ was augmented with an additional component, namely the next goal g, which assigns each state to a particular waypoint segment.

During the graph search, a transition from a state $\tilde{\mathbf{s}}^{d,r} = (\ldots, g_i)$ of the currently considered planning segment to a state $\tilde{\mathbf{s}}^{d,r'} = (\ldots, g_{i+1})$ of the next planning segment is performed if and only if $\tilde{\mathbf{s}}^{d,r'}$ is located inside the next waypoint g_i, i.e.,

$$\tilde{\mathbf{s}}^{d,r'} \in \tilde{\mathcal{S}}_{g_i} \,. \tag{5.11}$$

Figure 5.3 illustrates this concept for a simple two-dimensional state space and a task consisting of two waypoints.

The introduction of multiple-waypoint planning does not necessarily lead to increased computational complexity. Since all states are explicitly assigned to a waypoint segment at any given time, the addition of the goal dimension G to the lattice does not increase the dimensionality of the planning problem. Thus, the computational complexity does not depend on the actual number of waypoints but only on the length and intricacy of the final solution. However, special care must be taken when applying heuristic graph search methods. Since a heuristic cost function that only directs the search toward the final waypoint g_N would be too optimistic, it is required to use a specialized heuristic cost function that properly reflects the geometric relation of all waypoints $g_1, \ldots, g_N$. This is explained in detail in Section 5.4.1.

5.1.6 Multi-Resolution Trajectory with Hybrid Dimensionality

With the transitions defined in the previous sections, the multi-resolution trajectory ψ with hybrid dimensionality, which constitutes the solution of

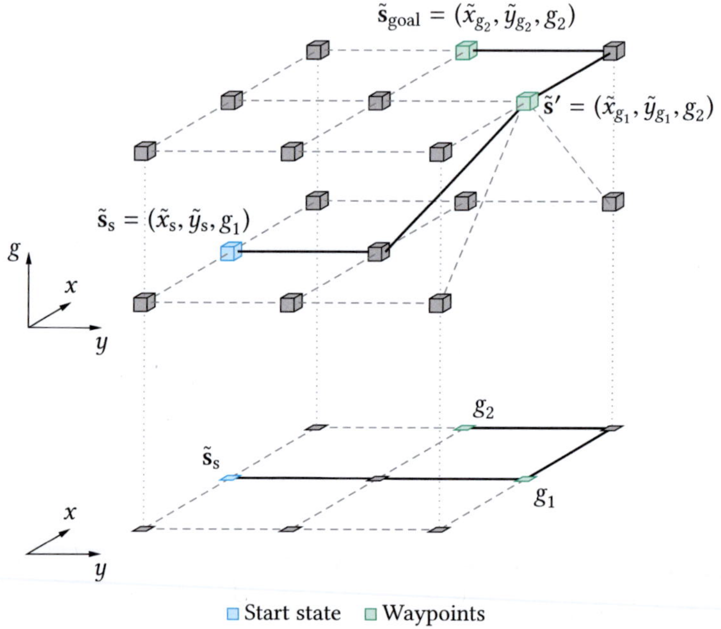

Figure 5.3: Graph for multiple-waypoint planning. The state $\tilde{s}'$ is the only connection between the g_1- and the g_2-layer in the search graph. Thus, the optimal path to the final goal g_2 is required to pass the intermediate waypoint g_1.

the proposed algorithm to the motion planning problem, is given by

$$\psi = \left(\tilde{\mathbf{s}}_0^{0,r_0}, \tilde{\mathbf{s}}_1^{0,r_1}, \dots, \tilde{\mathbf{s}}_i^{0,r_i}, \tilde{\mathbf{s}}_{i+1}^{1,r_{i+1}}, \dots, \tilde{\mathbf{s}}_j^{d_{\min}-1,r_j}, \tilde{\mathbf{s}}_{j+1}^{d_{\min},r_{j+1}}, \dots, \tilde{\mathbf{s}}_K^{d_{\min},r_K}\right), \tag{5.12}$$

where $\tilde{\mathbf{s}}_0^{0,r_0} = \tilde{\mathbf{s}}_\mathrm{s}^{0,r_0}$ is the start state and $\tilde{\mathbf{s}}_K^{d_{\min},r_K} \in \tilde{\mathcal{S}}_{g_N}$ is a state satisfying the goal condition. The trajectory ψ is a sequence of lattice points, i.e., discrete states, with decreasing dimensionality: A full-dimensional trajectory is available until state $\tilde{\mathbf{s}}_i^{0,r_i}$ with $\bar{t}(\tilde{\mathbf{s}}_i^{0,r_i}) \leq \tau_0$. From $\tilde{\mathbf{s}}_{i+1}^{1,r_{i+1}}$ on, with $\bar{t}(\tilde{\mathbf{s}}_{i+1}^{1,r_{i+1}}) > \tau_0$, the dimensionality is reduced by one level. Such a dimensionality reduction may occur several times until finally the transition to the lowest dimensionality is made at $\tilde{\mathbf{s}}_{j+1}^{d_{\min},r_{j+1}}$ where it remains till the last goal, which is reached after a total of K states. The resolution level r_k, $k = 0, \dots, K$ of each state in ψ may change at any time according to the criteria described in Section 5.1.4. In addition, a transition between different waypoint segments could occur anywhere along the trajectory and is fully independent of the dimensionality or resolution (see Section 5.1.5).

Since the state sequence in (5.12) originates from the sequential application of motion primitives, the motion primitives can be used to reconstruct the associated system controls $\mathbf{u}$. Due to potential discontinuities of state variables that have been dropped by the projection process, the system inputs are most meaningful for the beginning of the trajectory, which consists of full-dimensional states. It is precisely these early controls that are relevant for the imminent motion of the robot. The planning method proposed in this thesis provides an approximation to the solution of the general time-parametrized motion planning problem (1.13), for which no analytic solution exists for real-world applications in complex environments.

If a full-dimensional trajectory that is also optimal in the continuous sense is required, one has to resort to conventional numerical optimization techniques like sequential quadratic programming (SQP) or interior point (IP) methods in order to solve the optimal control problem (OCP)

(1.13). For these methods, which are inherently only able to find a local optimum, the lattice-based motion planning algorithm can provide valuable initial values, that are already a good guess for the optimal continuous solution since by design the graph search provides solutions that are globally optimal for the chosen resolution and set of motion primitives. The continuous optimization of the trajectory directly leads to the topic of trajectory smoothing, which is a research area of its own (for examples see [Bet98; Thr06b; Xu12]) and is therefore not discussed in this thesis. However, tests on a real robotic platform showed that the proposed hybrid-dimensional planning scheme is able to produce high-quality solutions without the need for further optimization. The planned trajectories can easily be tracked by conventional control methods in order to achieve a robust and accurate overall motion (see Section 6.4).

5.2 Finding the Shortest Path in the Graph

The approach that has been developed in this thesis converts the motion planning problem to the task of finding shortest paths in a graph. The construction of the search graph has been described in Section 5.1. Since the entire graph may be very large and extend to regions that are never explored during the actual graph search, the graph is not built up at the beginning of the planning cycle, but the relevant part is incrementally constructed as the search progresses and nodes get expanded.

5.2.1 Choice of Graph Search Algorithm

Finding shortest paths in a graph is an often-encountered task in robotics for which a wide range of established algorithms are readily available. An overview of heuristic graph search algorithms was provided in Section 2.6.3. There, it was pointed out that the backward search of incremental replanning algorithms is disadvantageous for the consideration of dynamic obstacles. Additionally, in the context of the proposed hybrid-dimensional planning scheme, it is not clear when a transition to another dimensionality level should be made because the thresholds τ_i are based on the duration measured from the start state, which is not known until finally reaching the start state during the backward search. On the other hand, anytime algorithms are able to quickly provide an initial (possibly suboptimal but perfectly valid) solution. This is why Anytime Repairing A* (ARA*) has been chosen for the computation of shortest paths through the lattice-based search graph in this thesis.

5.2.2 Anytime Repairing A*

This section highlights the key ideas of the ARA* algorithm and illustrates the basic concept. For the sake of clarity and in order to keep the notation short, the superscripts indicating the dimensionality and resolution level of a state $\tilde{s}^{d,r}$ are omitted in the following sections, and thus a state of the search graph is briefly denoted by $\tilde{s}$.

ARA* was proposed by Likhachev, Gordon, and Thrun in [Lik03a]. Similar to the regular A* algorithm, states that have yet to be expanded are collected in a priority queue called *OPEN*. They are sorted based on a key, also called f-value,

$$f(\tilde{s}) := g(\tilde{s}) + \epsilon\, h(\tilde{s})\,. \tag{5.13}$$

The state with minimum f-value is always expanded first and added to the *CLOSED* set, which keeps track of all expanded states. The f-value is computed by the sum of the accumulated cost $g(\tilde{s})$ for reaching $\tilde{s}$ and the estimated remaining cost to the goal, which is the heuristic cost $h(\tilde{s})$. The parameter ϵ is called the *heuristic inflation factor*. For $\epsilon = 1$, the f-value (5.13) is a lower bound for the cost of a path to the goal leading through $\tilde{s}$.

It is assumed that the heuristic $h(\tilde{s})$ is consistent, which means that

$$h(\tilde{s}) \le h(\tilde{s}') + c(\tilde{s}, \tilde{s}') \tag{5.14}$$

holds for any $\tilde{s}' \in \text{succ}(\tilde{s})$. Here, $c(\tilde{s}, \tilde{s}')$ denotes the cost for the transition from $\tilde{s}$ to $\tilde{s}'$. A consistent heuristic implies that it is also admissible, i.e., it never overestimates the actual cost to the goal [Pea84]. With an admissible heuristic, A* is guaranteed to find the optimal path when expanding the states in increasing order of f-values. A heuristic that is in addition consistent ensures that A* expands each state at most once during the search [Lik03a].

The basic idea of ARA* is to deliberately violate the consistency assumption (5.14) and perform the initial planning with an increased heuristic inflation factor $\epsilon > 1$ similar to Weighted A*. This way, the algorithm can quickly provide an initial—but possibly suboptimal—solution, which is then iteratively refined by lowering the inflation factor ϵ until the allocated computation time is spent or the optimal solution is found for $\epsilon = 1$. Since iteratively replanning from scratch is prohibitively expensive, ARA* makes extensive use of search results from previous iterations. For this purpose, the notion of *local inconsistency* is introduced [Lik03a]. A state $\tilde{s}$ is called locally inconsistent if its g-value $g(\tilde{s})$ has been lowered in the current iteration but the state $\tilde{s}$ itself is yet to be expanded. Before lowering $g(\tilde{s})$, the accumulated cost of a successor $\tilde{s}' \in \text{succ}(\tilde{s})$ is obtained by

$$g(\tilde{s}') = g(\tilde{s}) + c(\tilde{s}, \tilde{s}') \,. \tag{5.15}$$

However, after lowering $g(\tilde{s})$, equation (5.15) is no longer valid; instead, the decreased $g(\tilde{s})$ results in

$$g(\tilde{s}') > g(\tilde{s}) + c(\tilde{s}, \tilde{s}') \,, \tag{5.16}$$

and thus $\tilde{s}$ is said to be locally inconsistent. The inconsistency is corrected as soon as $\tilde{s}$ is expanded and its successors' costs are updated according to (5.15), which in turn renders their own successors locally inconsistent, thus propagating the local inconsistency to the children of $\tilde{s}$.

While regular A* guarantees that no state is expanded more than once due to the consistent heuristic, this is no longer true when planning with an increased heuristic inflation factor $\epsilon > 1$. This is why, besides the *OPEN* set, which contains all states that are yet to be expanded, and the *CLOSED* set, which contains all states that have already been expanded, ARA* introduces an additional set, namely *INCONS*, which contains all states that have been found to be inconsistent but are not a member of *OPEN*. This inconsistency will be corrected by re-expanding those states in the next ARA* iteration. For this purpose, at the beginning of each ARA* iteration, the heuristic inflation factor ϵ is reduced and all states of *INCONS* are moved to *OPEN*. Due to the changed ϵ, the f-value (5.13) of all states in *OPEN* needs to be re-evaluated in order to update the sorting of the priority queue which consists of the states in *OPEN*. The set of expanded states from the previous iteration, *CLOSED*, is cleared so that it can record expanded states of the current ARA* iteration. With this preparation, the actual graph search is performed in a similar manner to the regular A* algorithm. However, as soon as a state that has been expanded before is revisited and its g-value is lowered, it is marked as inconsistent by adding it to *INCONS*. The algorithmic details of ARA* are cited in Appendix B, Algorithm 2 on page 211 for reference.

The solution of an ARA* iteration is guaranteed to be at most suboptimal by a factor of ϵ [Lik03b]. In fact, an even more accurate suboptimality bound can be obtained by the ratio of $g(\tilde{s}_{\text{goal}})$, i.e., the accumulated cost to the goal from the previous iteration, and the non-inflated minimum f-value

$$\underline{f} = \min_{\tilde{s} \in OPEN \cup INCONS} \left(g(\tilde{s}) + h(\tilde{s}) \right) \tag{5.17}$$

of all inconsistent states [Lik08]. Since both ϵ and the ratio $g(\tilde{s}_{\text{goal}})/\underline{f}$ are valid suboptimality bounds, their minimum

$$\epsilon' = \min\left(\epsilon, \frac{g(\tilde{s}_{\text{goal}})}{\min_{\tilde{s} \in OPEN \cup INCONS}\left(g(\tilde{s}) + h(\tilde{s})\right)}\right) \tag{5.18}$$

is also a valid suboptimality bound [Lik08]. This suboptimality bound provides a valuable assessment of the optimality of the planned motion if the computation time budget is exhausted before the provably optimal solution for $\epsilon = 1$ is found by ARA*.

The ARA* algorithm achieves a substantial speed-up over iteratively planning from scratch in a Weighted A* manner. For example, the authors of [Lik03a] observed a 140-fold speed-up when planning a path for a robotic arm with six degrees of freedom. They found the overhead of ARA* compared to optimal A* search to be about 4 % when planning with an initial $\epsilon = 3$ and successively reducing ϵ in steps of 0.02 until the optimal solution was found for $\epsilon = 1$. The results for application of ARA* for motion planning in multi-resolution state × time × goal lattices with hybrid dimensionality in the context of this thesis are presented in Chapter 6.

5.3 Costs

The computation of the g-value (5.15) is based on the cost $c(\tilde{s}, \tilde{s}')$ of the transition from $\tilde{s}$ to $\tilde{s}'$. Various quantities may contribute to that cost: The most obvious and most frequently used component is the distance traveled from the start state, but also the elapsed travel time since the start may be of interest. In addition, the traversed terrain may be incorporated into the cost. This can be done either by applying a penalty for traversing certain areas or by computing the terrain's influence on the robot (e.g., roll and pitch angle [Lin08]). For autonomous passenger vehicles, maximizing comfort (e.g., by minimizing jerk) is generally a high priority [Lev11].

Besides the traveled distance and time, the implementation of the planning algorithm for the mobile robot from Section 3.4 also strives to minimize the collision risk caused by static and dynamic obstacles or terrain-specific features (see Chapter 4) during the execution of a motion primitive

$$m = (\tilde{\mathbf{s}}_0, \tilde{\mathbf{s}}_e, \mathbf{u}, \mathcal{T}, \Delta t_m) . \tag{5.19}$$

The length of the motion primitive is defined by the length of the 2D path corresponding to its trajectory $\mathcal{T}$ and is denoted by $l(m)$. For a state $\tilde{\mathbf{s}}$, the traveled distance since the start state $\tilde{\mathbf{s}}_s$ is the sum of the individual lengths of all motion primitives that led to $\tilde{\mathbf{s}}$ and is denoted by $\bar{l}(\tilde{\mathbf{s}})$. The time for executing a motion primitive is tantamount to its duration Δt_m, which may also be written as $t(m)$ for consistency. The cumulative time for reaching a state $\tilde{\mathbf{s}}$ is denoted by $\bar{t}(\tilde{\mathbf{s}})$. The collision risk $p_{coll}(m)$ for executing the motion primitive is obtained by (4.13), and the overall collision risk for executing the trajectory from the start state $\tilde{\mathbf{s}}_s$ to a given state $\tilde{\mathbf{s}}$ is denoted by $\bar{p}_{coll}(\tilde{\mathbf{s}})$.

With these quantities the cost of the trajectory from the start state $\tilde{\mathbf{s}}_s$ to a state $\tilde{\mathbf{s}}$ can be defined as

$$g(\tilde{\mathbf{s}}) := \bar{l}(\tilde{\mathbf{s}}) + \eta_t \bar{t}(\tilde{\mathbf{s}}) + \eta_r \bar{p}_{coll}(\tilde{\mathbf{s}}) \tag{5.20}$$

with $\eta_t > 0$ and $\eta_r \geq 0$ being weighting factors to adapt the cost function to suit the particular requirements of a scenario [Pet13c]. The factor η_r controls the acceptable detour for finding a path with lower collision risk. The cost for reaching a successor $\tilde{\mathbf{s}}' \in \mathrm{succ}(\tilde{\mathbf{s}})$ is computed using the quantities of the corresponding motion primitive:

$$\begin{aligned} g(\tilde{\mathbf{s}}') &= \bar{l}(\tilde{\mathbf{s}}') + \eta_t \bar{t}(\tilde{\mathbf{s}}') + \eta_r \bar{p}_{coll}(\tilde{\mathbf{s}}') \\ &= \bar{l}(\tilde{\mathbf{s}}) + l(m) + \eta_t\big(\bar{t}(\tilde{\mathbf{s}}) + t(m)\big) \\ &\quad + \eta_r\Big(1 - \big(1 - \bar{p}_{coll}(\tilde{\mathbf{s}})\big)\big(1 - p_{coll}(m)\big)\Big) . \end{aligned} \tag{5.21}$$

This formulation of the successors' g-value ensures that the cost for expanded nodes is monotonically increasing along the path, i.e., $c(\tilde{\mathbf{s}}, \tilde{\mathbf{s}}') > 0$ for all $\tilde{\mathbf{s}}' \in \mathrm{succ}(\tilde{\mathbf{s}})$ (see Proposition 1, p. 203). This property is required for the application of ARA* [Lik03b]. A graph with arbitrary edge costs would require the application of more general algorithms, like the Bellman-Ford algorithm, for finding the shortest path in the graph [Cor09].

The cost formulation (5.21) may be extended to reflect further preferences. For example, driving backwards may be penalized by multiplying $l(m)$ with a factor $\eta_\mathrm{b} > 1$ whenever m causes backward motion of the robot.

A graph search algorithm finds the least-cost path to the goal. This is equivalent to solving the optimization problem

$$g^*(\tilde{\mathbf{s}}_\mathrm{goal}) = \min_{\tilde{\mathbf{s}}_\mathrm{goal} \in \tilde{\mathcal{S}}_{g_N}} g(\tilde{\mathbf{s}}_\mathrm{goal}) \,. \tag{5.22}$$

where $g(\tilde{\mathbf{s}}_\mathrm{goal})$ depends on the trajectory/path to $\tilde{\mathbf{s}}_\mathrm{goal}$, which is constructed by the motion primitive concatenation according to Section 5.1. The accumulated costs g in (5.22) are the graph search counterpart to the cost functional J in (1.13).

5.4 Heuristics

Heuristics provide an estimate of the remaining costs to the goal. Since the state × time lattice is a metric space, heuristic search algorithms are well suited for mobile robot motion planning because the inherent structure of the search space allows a relatively straightforward estimation of the remaining costs. For fast computation of the heuristic costs, it is a common approach to solve a surrogate relaxed problem [Rus95]. In the context of robot motion planning, such a relaxation may be obtained by disregarding of obstacles in the environment or considering of kinematical paths instead of the full-dimensional trajectory.

The most pragmatic approach is the computation of the Euclidean distance to the goal; however, more advanced strategies may be employed, like Dubins curves [Dub57], which allow the analytical computation of a vehicle's shortest path by concatenation of circular arcs and straight line segments. They have been extended to include backward motion (Reeds-Shepp curves) [Ree90] and to consider continuous curvature of the path [Sch98]. All these heuristics assume a mostly free configuration space, which is generally not true for unstructured environments. The more obstacles there are in the environment, the less informative is such a heuristic. This is why efforts have been made to incorporate the structure of the free space into the heuristic cost. For example, an approach that constructs a Voronoi diagram of the free space and derives a heuristic cost function from this diagram is investigated in [Zie08]; however, the authors did not prove that their proposed heuristic is admissible or even consistent. In addition, the Voronoi diagram needs to be recomputed from scratch each time the map has changed, which may be prohibitively expensive for large maps. In pursuit of more accurate, and thus more informative, heuristics, the authors of [Kne06] compute a heuristic look-up table (HLUT) which contains the costs for all admissible motions in a four-dimensional state lattice without consideration of obstacles; of course, this can only be done for a spatially confined region due to memory limitations for storing the HLUT. This approach has been combined with an obstacle-aware 2D grid search in [Fer08b] by using the maximum of the individual heuristics. This strategy proved its worth in practice and similar methods have been used by team Stanford's *Junior* [Dol10] or team *AnnieWAY* [Kam08] during the DARPA Urban Challenge.

The heuristics used in this thesis build on these established concepts but are substantially extended to fit the requirements for multiple-waypoint planning (Section 5.4.1). Additionally, an improved 2D obstacle-aware heuristic is developed in Section 5.4.2, which is able to provide more precise estimates on the remaining distance than the Dijkstra-based algorithm proposed in [Lik09].

Following the definition of the cost function (5.20), the heuristic costs comprise a spatial and a temporal component. The collision risk, however, is not considered since it is impossible to draw conclusions on the expected minimum collision risk without time-consuming examination of the actual environment. Therefore, the minimum expected collision risk is set to zero.

Unless otherwise stated, the remainder of this thesis assumes disk-shaped goal regions, also called waypoints, which are specified by their center (x_g, y_g) and their radius R_g. The target orientation of the robot when reaching a goal is left unspecified as an optimization parameter. Thus, the set of goal states is given by

$$\tilde{\mathcal{S}}_g := \left\{ \tilde{\mathbf{s}} \,\middle|\, (\tilde{x} - x_g)^2 + (\tilde{y} - y_g)^2 \leq R_g^2 \right\}. \tag{5.23}$$

A lower bound $\mathfrak{d}(\tilde{\mathbf{s}})$ on the distance from a state $\tilde{\mathbf{s}} = (\tilde{x}, \tilde{y}, \ldots, g)$ to the goal g with its associated set of goal states $\tilde{\mathcal{S}}_g$ can be obtained by the Euclidean distance

$$\mathfrak{d}(\tilde{\mathbf{s}}) := \max\left(0, \sqrt{(\tilde{x} - x_g)^2 + (\tilde{y} - y_g)^2} - R_g\right), \tag{5.24}$$

which is illustrated in Figure 5.4. On the basis of this minimum distance, the temporal component of the heuristic cost function can be defined as follows. It is the remaining time $\mathfrak{t}(\tilde{\mathbf{s}})$ that is at least necessary to reach the goal:

$$\mathfrak{t}(\tilde{\mathbf{s}}) := \frac{\mathfrak{d}(\tilde{\mathbf{s}})}{\tilde{v}_{\max}}, \tag{5.25}$$

where $\tilde{v}_{\max} = \max_{\tilde{v} \in V} |\tilde{v}|$ with V defined according to (3.28).

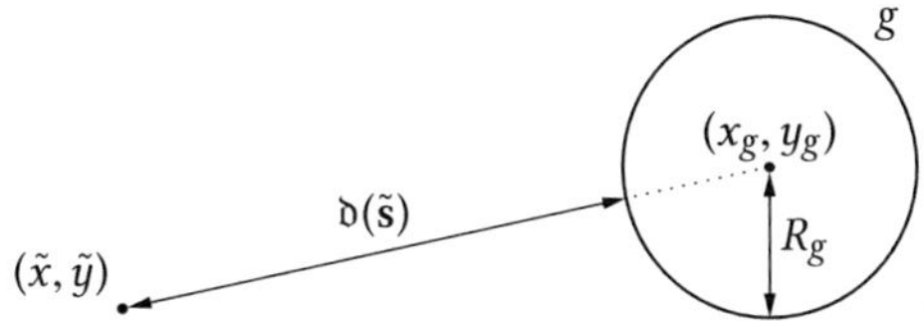

Figure 5.4: Heuristics for one waypoint.

With the spatial and temporal components (5.24) and (5.25), the heuristic cost function for single-waypoint planning is finally given by

$$h(\tilde{\mathrm{s}}) := \mathfrak{d}(\tilde{\mathrm{s}}) + \eta_{\mathrm{t}} \mathfrak{t}(\tilde{\mathrm{s}}) = \left(1 + \frac{\eta_{\mathrm{t}}}{\tilde{v}_{\max}}\right) \mathfrak{d}(\tilde{\mathrm{s}}) \,. \tag{5.26}$$

The heuristic cost function (5.26) is both admissible and consistent. As (5.26) is a special case of the multi-waypoint heuristics derived in the next section, the same proof applies (see Proposition 2, p. 204).

5.4.1 Heuristics for Multiple Waypoints

When planning for multiple waypoints simultaneously (see Sections 3.8 and 5.1.5), specially tailored heuristics are required in order to reflect the structure of the state × time × goal lattice with its corresponding search graph. The heuristic cost function needs to guide the state expansion during the graph search along the chain of waypoints. For this purpose, a generalized formulation has been developed to obtain a lower bound on the remaining costs for a task consisting of N waypoints.

Similar to the single-waypoint case, the heuristic cost function consists of a spatial and a temporal part according to (5.26). Thus, it is sufficient to derive a formula for the lower bound $\mathfrak{d}(\tilde{\mathrm{s}})$ on the remaining distance to the last waypoint along a path that leads through all intermediate waypoints. For a state $\tilde{\mathrm{s}} = (\tilde{x}, \tilde{y}, \ldots, g_i)$ belonging to the planning segment to waypoint g_i, the lower bound $\mathfrak{d}(\tilde{\mathrm{s}})$ can be split into two parts, namely the minimum distance $\mathfrak{d}_i(\tilde{\mathrm{s}})$ to the next waypoint g_i and the remaining minimum distance from g_i to g_N via $g_{i+1}, \ldots, g_{N-1}$:

$$\mathfrak{d}(\tilde{\mathrm{s}}) := \mathfrak{d}_i(\tilde{\mathrm{s}}) + \sum_{j=i+1}^{N} \mathfrak{w}_j \tag{5.27}$$

with $\mathfrak{d}_i(\tilde{\mathrm{s}})$ defined analogously to (5.24),

$$\mathfrak{d}_i(\tilde{\mathrm{s}}) := \max\left(0, \sqrt{(\tilde{x} - x_{g_i})^2 + (\tilde{y} - y_{g_i})^2} - R_{g_i}\right), \tag{5.28}$$

and the inter-waypoint distance

$$\mathfrak{w}_j := \max\left(0, \sqrt{(x_{g_j} - x_{g_{j-1}})^2 + (y_{g_j} - y_{g_{j-1}})^2} - R_{g_{j-1}} - R_{g_j}\right). \tag{5.29}$$

The computation of the heuristic distance (5.27) is illustrated in Figure 5.5 for a sample task consisting of three waypoints. The final spatiotemporal heuristic cost function $h(\tilde{\mathrm{s}})$ can be computed by substituting (5.27) into definition (5.26). For $N = 1$, i.e., single-waypoint planning, the multi-waypoint heuristic distance (5.27) is equivalent to its single-waypoint counterpart (5.24). The proposed heuristic multi-waypoint cost function is consistent and thus also admissible (see Proposition 2, p. 204, for the proof).

5.4.2 Obstacle-Aware 2D Heuristics

The heuristic cost function proposed in the previous section allows a very fast computation of a lower bound on the remaining distance to the goal. However, as those heuristics ignore all obstacles and thus the topology of the environment, they may vastly underestimate the remaining distance and may thus be not as informative as it is desirable. For this reason,

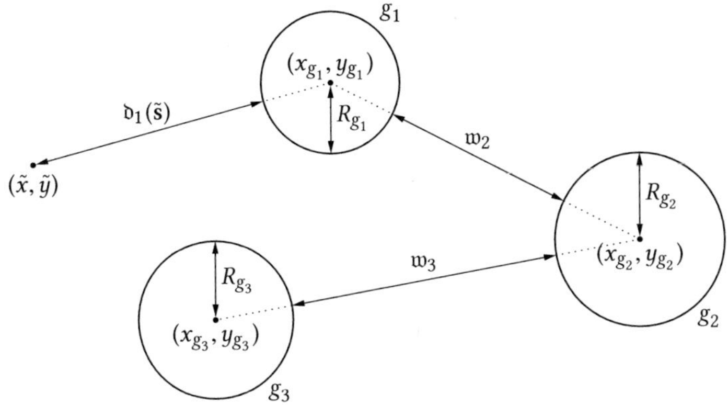

Figure 5.5: Heuristics for multiple waypoints.

additional obstacle-aware heuristics are often employed [Fer08b]. They commonly consist of a heuristic cost map that is obtained by a simple grid search. For example, the details of the heuristics used in [Fer08b] are provided in [Lik09]: The authors apply a Dijkstra-based search to an eight-connected grid to determine the shortest paths from the goal to each cell of the map. Horizontal and vertical motions incur a cost of 1 and moving to a diagonally connected cell incurs a cost of $\sqrt{2}$. Since only directions of motion that are a multiple of $\pi/4$ are allowed in an eight-connected grid, the least-cost paths generally overestimate the true distance under an Euclidean metric (except for purely straight or diagonal paths). This error attains its maximum value for paths with straight and diagonal portions of equal length (see Figure 5.6). In this case, the grid search may overestimate the Euclidean distance by a factor of

$$\nu := \frac{1}{\cos(\pi/8)} \approx 1.0824\,, \tag{5.30}$$

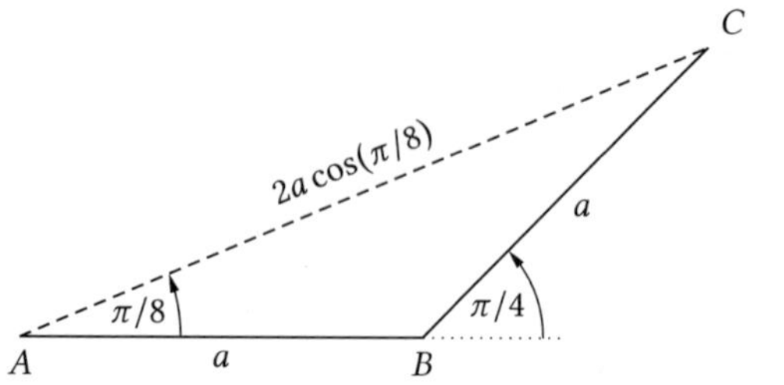

Figure 5.6: Worst-case scenario for overestimation of heuristic distance in an eight-connected grid: The shortest path from A to C computed by the Dijkstra-like search in an eight-connected grid [Lik09] has a total length of $2a$ (solid lines); however, the Euclidean distance from A to C is only $2a\cos(\pi/8)$ (dashed line). Thus, the shortest path in the eight-connected grid overestimates the Euclidean distance by a factor of $1/\cos(\pi/8)$.

or roughly 8 %. In order to guarantee that the computed heuristic cost function is still admissible, the authors divide all costs by ν. As a consequence, the computed heuristic costs for all cases that do not follow the worst-case scenario from Figure 5.6 are now in turn underestimated. Especially for mostly straight or diagonal paths, the remaining distance may be underestimated by up to the factor ν, which makes the heuristic cost function less informative. This results in more states to be unnecessarily expanded during the actual ARA* graph search.

In order to avoid the disadvantages mentioned above, a new obstacle-aware 2D heuristic has been developed in this thesis. It is also based on a Dijkstra-like grid search but uses a different cost function. For the derivation of the improved heuristic, the cost computation during the search in an eight-connected grid for the established heuristic from [Lik09] is restated in closed-form in the following. In the remainder of this section, a state $\tilde{\mathbf{s}}_i = (\tilde{x}_i, \tilde{y}_i, \dots)$ is interpreted as the vector $[\tilde{x}_i \;\; \tilde{y}_i]^\top$ to be able to draw on some tools of linear algebra. Without loss of generality, it is assumed that the optimal path from a state $\tilde{\mathbf{s}}_0$ to a state $\tilde{\mathbf{s}}_N$ leads through

$N - 1$ critical points (i.e., states) that are determined by the obstacles in the environment. The derivation of the improved heuristic assumes that the critical points are known; however, this is not necessary for the actual implementation.

Figure 5.7a shows an example for the optimal 2D path from $\tilde{\mathbf{s}}_0$ to $\tilde{\mathbf{s}}_2$ via the critical point $\tilde{\mathbf{s}}_1$. The dashed line depicts the shortest path from $\tilde{\mathbf{s}}_0$ to $\tilde{\mathbf{s}}_2$ via $\tilde{\mathbf{s}}_1$ under the Euclidean metric. Its length is obtained by the sum of the length of its straight-line segments $\mathbf{p}_i := \tilde{\mathbf{s}}_i - \tilde{\mathbf{s}}_{i-1}$,

$$\mathfrak{d}_{\mathrm{Eucl}} := \sum_{i=1}^{N} \|\mathbf{p}_i\| = \sum_{i=1}^{N} \|\tilde{\mathbf{s}}_i - \tilde{\mathbf{s}}_{i-1}\| , \tag{5.31}$$

where N is the number of straight-line segments for $N - 1$ critical points, and $\|\cdot\|$ denotes the Euclidean norm. With this notation, the length of the shortest path computed by the Dijkstra-like grid search in [Lik09] can be computed by

$$\mathfrak{d}_{\mathrm{Lik}} := \frac{1}{\nu} \sum_{i=1}^{N} \left(\max\left(|p_{x,i}|, |p_{y,i}|\right) + \left(\sqrt{2} - 1\right) \min\left(|p_{x,i}|, |p_{y,i}|\right) \right) . \tag{5.32}$$

This corresponds to the length of the solid lines in Figure 5.7a (with the additional correction factor ν). To shorten notation, the vector

$$\mathbf{r}_i := \begin{bmatrix} \max(|p_{x,i}|, |p_{y,i}|) \\ \min(|p_{x,i}|, |p_{y,i}|) \end{bmatrix} \tag{5.33}$$

is introduced, which mirrors and/or rotates $\mathbf{p}_i$ such that its largest absolute component points in x-direction and its minimum absolute component points in y-direction. This transformation does not change the length of

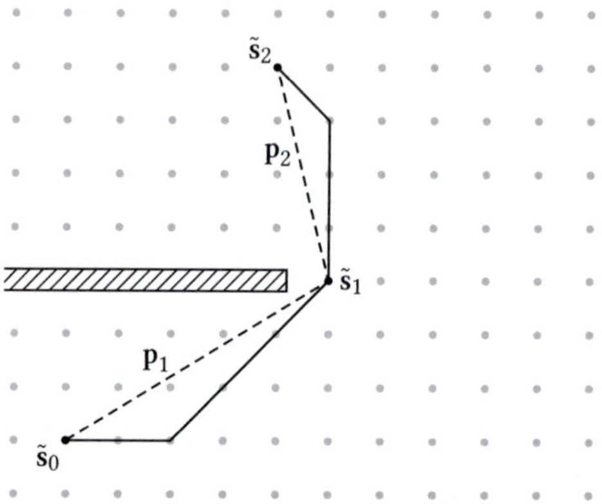

(a) Initial setup: Euclidean distance $\mathfrak{d}_{\text{Eucl}}$ (dashed lines) and heuristic distance function $\nu\mathfrak{d}_{\text{Lik}}$ (solid lines).

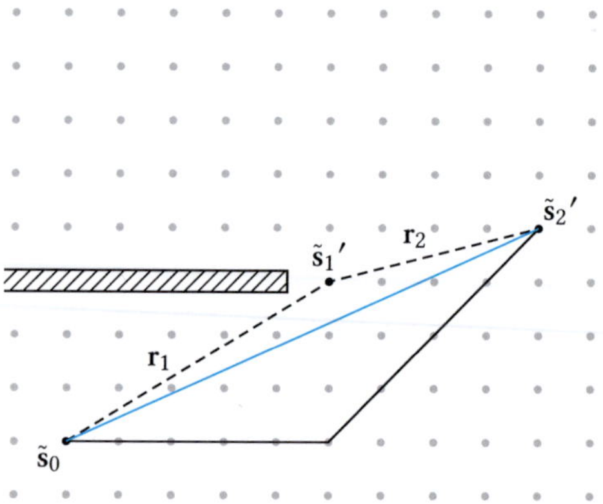

(b) Scenario after application of transformation (5.33): Euclidean distance $\mathfrak{d}_{\text{Eucl}}$ (dashed lines), heuristic distance function $\nu\mathfrak{d}_{\text{Lik}}$ (solid lines), and novel improved heuristic distance function $\mathfrak{d}_{\text{obst}}$ (blue line).

Figure 5.7: Obstacle-aware 2D heuristic distance function computation: sample scenario with one critical point.

the vector, i.e., $\|\mathbf{r}_i\| = \|\mathbf{p}_i\|$. With (5.33), (5.32) can be rewritten as

$$\begin{aligned} \mathfrak{d}_{\text{Lik}} &= \frac{1}{v}\sum_{i=1}^{N}\left(r_{x,i} + (\sqrt{2}-1)r_{y,i}\right) \\ &= \frac{1}{v}\sum_{i=1}^{N}\begin{bmatrix}1 \\ \sqrt{2}-1\end{bmatrix}^{\top}\mathbf{r}_i\,. \end{aligned} \tag{5.34}$$

This still corresponds to the solid line in Figure 5.7a (leaving aside the factor v). Instead of adding the length of each segment, (5.34) can also be written as

$$\mathfrak{d}_{\text{Lik}} = \frac{1}{v}\begin{bmatrix}1 \\ \sqrt{2}-1\end{bmatrix}^{\top}\sum_{i=1}^{N}\mathbf{r}_i\,, \tag{5.35}$$

which first sums all vectors $\mathbf{r}_i$ before performing the distance computation. This is illustrated by the black solid lines in Figure 5.7b, whose length is identical to the length of the solid lines in Figure 5.7a. On the basis of (5.35), a new improved obstacle-aware 2D heuristic distance function is proposed in this thesis. It exploits definition (5.33) and is defined by

$$\mathfrak{d}_{\text{obst}} := \left\|\sum_{i=1}^{N}\mathbf{r}_i\right\|. \tag{5.36}$$

This corresponds to the length of the vector resulting from the summation of all vectors $\mathbf{r}_i$, which is visualized by the blue line in Figure 5.7b. The new heuristic (5.36) is admissible since it never overestimates the Euclidean distance (5.31). This becomes clear from looking at Figure 5.7b and can be proved by the application of the triangle inequality:

$$\mathfrak{d}_{\text{obst}} = \left\|\sum_{i=1}^{N}\mathbf{r}_i\right\| \le \sum_{i=1}^{N}\|\mathbf{r}_i\| = \sum_{i=1}^{N}\|\mathbf{p}_i\| = \mathfrak{d}_{\text{Eucl}}\,. \tag{5.37}$$

In addition, the new heuristic $\mathfrak{d}_{\text{obst}}$ also dominates the heuristic $\mathfrak{d}_{\text{Lik}}$, i.e., it is always more or equally informative ($\mathfrak{d}_{\text{obst}} \geq \mathfrak{d}_{\text{Lik}}$). To prove this property, the vector

$$\mathbf{a} := \frac{1}{v}\begin{bmatrix} 1 \\ \sqrt{2}-1 \end{bmatrix} \tag{5.38}$$

is introduced. Its norm can be computed with the help of the half-angle formula for the cosine,

$$\cos(\alpha/2) = \sqrt{\frac{1+\cos(\alpha)}{2}}\,, \tag{5.39}$$

see [Olv10], to give

$$\|\mathbf{a}\| = \left\| \cos(\pi/8)\begin{bmatrix} 1 \\ \sqrt{2}-1 \end{bmatrix} \right\| = \sqrt{\frac{1+\cos(\pi/4)}{2}}\sqrt{4-2\sqrt{2}} = 1\,. \tag{5.40}$$

The dominance of $\mathfrak{d}_{\text{obst}}$ over $\mathfrak{d}_{\text{Lik}}$ can now be proved by exploiting the Cauchy-Schwarz inequality (CSI) [Olv10]:

$$\begin{aligned} \mathfrak{d}_{\text{Lik}} &= \mathbf{a}^{\top}\sum_{i=1}^{N}\mathbf{r}_i = \left\| \mathbf{a}^{\top}\sum_{i=1}^{N}\mathbf{r}_i \right\| \\ &\overset{\text{CSI}}{\leq} \|\mathbf{a}\| \left\| \sum_{i=1}^{N}\mathbf{r}_i \right\| \overset{(5.40)}{=} \left\| \sum_{i=1}^{N}\mathbf{r}_i \right\| = \mathfrak{d}_{\text{obst}}\,. \end{aligned} \tag{5.41}$$

The new heuristic $\mathfrak{d}_{\text{obst}}$ is also implemented in terms of a Dijkstra-like search in an eight-connected grid. Its cost, however, is computed differently from the method proposed in [Lik09]. While the latter assigns the cost in an incremental fashion by adding $1/v$ for vertical and horizontal motion and $\sqrt{2}/v$ for diagonal motion, the new heuristic $\mathfrak{d}_{\text{obst}}$ keeps track of the sum $\sum_{i=1}^{N}\mathbf{r}_i$ by using $\mathbf{r}_i = \left[\begin{smallmatrix}1\\0\end{smallmatrix}\right]$ for horizontally or vertically adjacent cells and $\mathbf{r}_i = \left[\begin{smallmatrix}1\\1\end{smallmatrix}\right]$ for diagonally adjacent cells (corresponding to definition

(5.33)). With this sum, the heuristic distance $\mathfrak{d}_{\text{obst}}$ can then be computed by (5.36). This formulation does not require explicit knowledge of the actual critical points, which were used in the derivation of the heuristic. It is sufficient to know the sum of the segments $\mathbf{r}_i$ that connect those points.

Figures 5.8 to 5.10 show a comparison of $\mathfrak{d}_{\text{Lik}}$ and $\mathfrak{d}_{\text{obst}}$. At a first glance, both heuristics seem to be identical (Figure 5.8). Therefore, Figure 5.9 visualizes the same data with a more expressive color map, which better illustrates the propagation of the "wave fronts," i.e., cells with identical heuristic distance. It can be observed that the propagation of $\mathfrak{d}_{\text{obst}}$ (Figure 5.9b) is more circular and thus more realistic than the propagation of $\mathfrak{d}_{\text{Lik}}$ (Figure 5.9a). Figure 5.10 shows the quantitative difference between the two heuristics: The absolute difference, $\mathfrak{d}_{\text{obst}} - \mathfrak{d}_{\text{Lik}}$, increases with increasing distance from the goal (Figure 5.10a) while the relative difference, $(\mathfrak{d}_{\text{obst}} - \mathfrak{d}_{\text{Lik}})/\mathfrak{d}_{\text{obst}}$, is especially large in the vicinity of the goal (Figure 5.10b). The maximum difference is roughly 8 %, which corresponds to $1 - 1/\nu$. Furthermore, it can be seen very clearly that the difference is particularly prominent in straight and diagonal directions since the application of the correction factor ν is most misleading for these directions.

For multiple-waypoint planning, the obstacle-aware 2D heuristics need to be computed for each waypoint segment individually. In principle, it is possible to compute the obstacle-aware 2D heuristics for each cell of the map. However, since the map may be very large, this would potentially involve computing the heuristics for large regions of the map that are never visited during the actual graph search. For this reason, restricting the computation of the heuristics to a threshold $\bar{\mathfrak{d}}_{\text{obst}}$ of about 100 m proved reasonable in practice. For all other cells, for which an obstacle-aware 2D heuristic is not available, the already introduced obstacle-unaware heuristic distance (5.27) is used. Whenever a transition is made between these two heuristics during the graph search, it may happen that the consistency condition for a heuristic cost function, see (A.4), is violated. Fortunately,

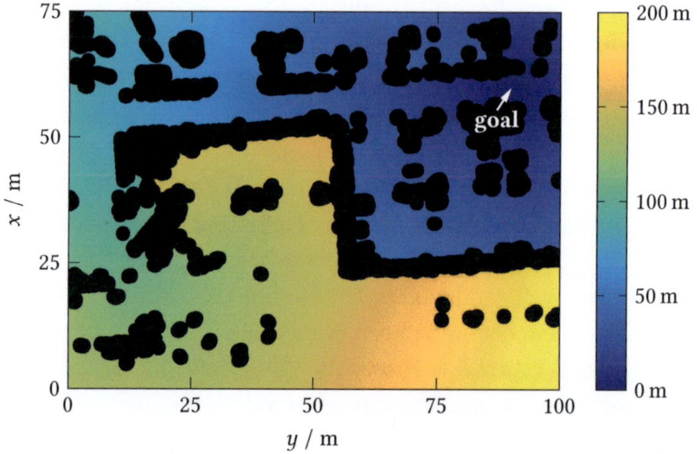

(a) Heuristic distance $\mathfrak{d}_{\mathrm{Lik}}$.

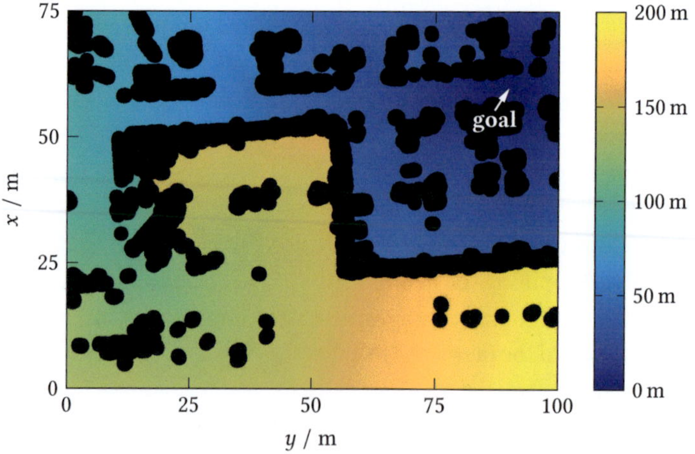

(b) Heuristic distance $\mathfrak{d}_{\mathrm{obst}}$.

Figure 5.8: Comparison of 2D obstacle-aware heuristic distances $\mathfrak{d}_{\mathrm{Lik}}$ and $\mathfrak{d}_{\mathrm{obst}}$. Obstacles are shown in black.

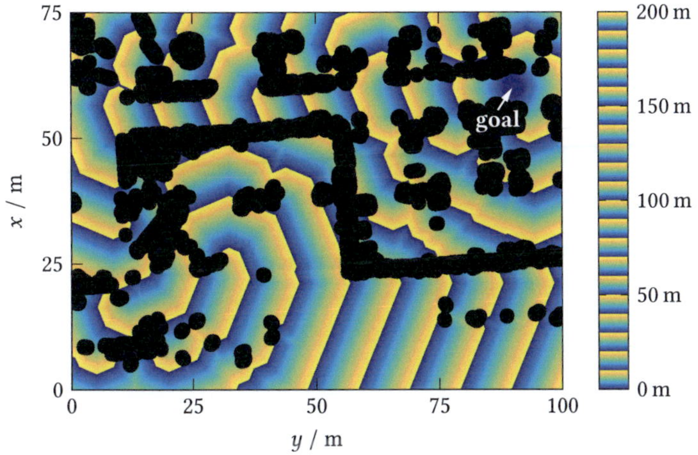

(a) Heuristic distance $\mathfrak{d}_{\mathrm{Lik}}$.

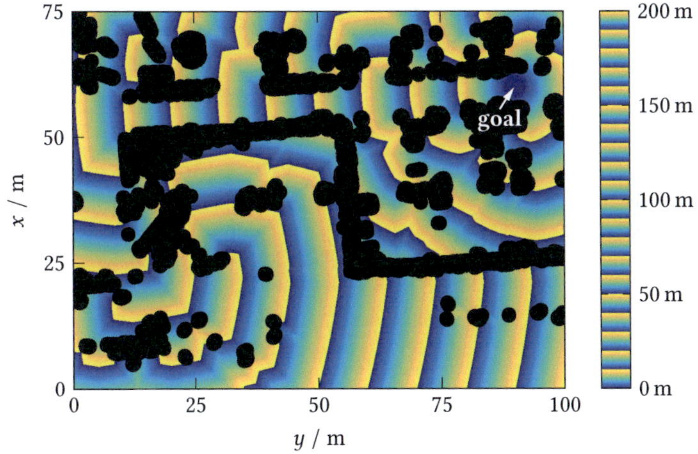

(b) Heuristic distance $\mathfrak{d}_{\mathrm{obst}}$.

Figure 5.9: Comparison of 2D obstacle-aware heuristic distances $\mathfrak{d}_{\mathrm{Lik}}$ and $\mathfrak{d}_{\mathrm{obst}}$: "wave front" propagation. Obstacles are shown in black.

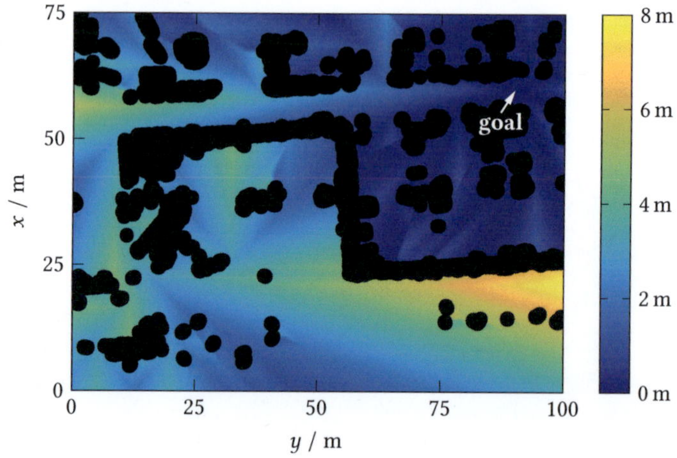

(a) Absolute difference of heuristic distances: $\mathfrak{d}_{\mathrm{obst}} - \mathfrak{d}_{\mathrm{Lik}}$.

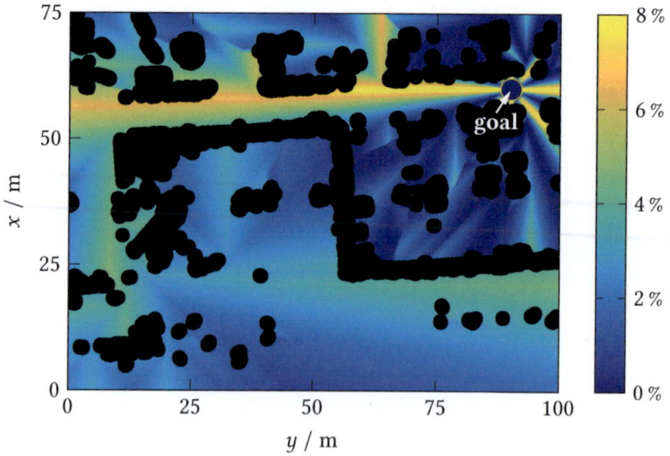

(b) Relative difference of heuristic distances: $(\mathfrak{d}_{\mathrm{obst}} - \mathfrak{d}_{\mathrm{Lik}})/\mathfrak{d}_{\mathrm{obst}}$.

Figure 5.10: Comparison of 2D obstacle-aware heuristic distances $\mathfrak{d}_{\mathrm{Lik}}$ and $\mathfrak{d}_{\mathrm{obst}}$: difference. Obstacles are shown in black.

a consistent heuristic h can be easily constructed from any admissible heuristic h' by application of the pathmax equation [Rus95],

$$f(\tilde{s}') = \max\left(f(\tilde{s}),\ g(\tilde{s}') + h'(\tilde{s}')\right) \tag{5.42}$$

or, alternatively,

$$h(\tilde{s}') = \max\left(h(\tilde{s}) + \underbrace{g(\tilde{s}) - g(\tilde{s}')}_{c(\tilde{s},\tilde{s}')},\ h'(\tilde{s}')\right) \tag{5.43}$$

with $\tilde{s}' \in \mathrm{succ}(\tilde{s})$ and $h'(\tilde{s}')$ being the tentative heuristic for the successor state $\tilde{s}'$. The pathmax equation is therefore used to ensure the consistency of the overall heuristic cost function.

5.5 Complexity

The main contribution of this thesis is the construction of a search graph based on a state × time × goal lattice structure for multi-resolution, multi-waypoint motion planning with hybrid-dimensionality (see Chapter 3). The proposed approach converts the complex motion planning problem to a regular search for a shortest path in a graph (Chapter 5). This allows the application of many well-established graph search algorithms—each one having its advantages and disadvantages as was discussed in Sections 2.6.3 and 5.2. Consequently, the achievable performance of the overall motion planning may vary depending on the selected algorithm and the particular scenario.

For the implementation in the context of this thesis a heuristic graph search method, ARA*, has been chosen. Thus, the performance of the planning does strongly depend on the quality of the utilized heuristics. If a constant heuristic cost function $h = 0$ was used, i.e., if the heuristic did not provide any information on the remaining cost (worst case), the search would degenerate to a breadth-first search with complexity exponential in the length of the solution. On the other hand, using a perfect heuristic which exactly matches the true remaining distance would result in a complexity linear in the length of the solution, since only states along the shortest path would ever be expanded [Pea84]. Of course, such a perfect heuristic is generally not available since its computation would be tantamount to solving the original planning problem. Thus, in reality, the performance is somewhere in between, and it is generally better the more informative the employed heuristic is.

In the context of mobile robot motion planning, suitable heuristics have been defined in Section 5.4. Since the informativeness of these heuristics depends on the structure of the respective environments and waypoint setups, a universal statement on the theoretical complexity cannot be made and would be rather unrewarding for the assessment of the real-world performance of the planning algorithm. For this reason, actual computation times observed during simulations and in the application for the real-world demonstrator are provided together with the results in Chapter 6 for a variety of scenarios. The results also evaluate the algorithm's susceptibility to changes in the parameters like the temporal thresholds τ_i.

Since the proposed algorithm reduces the motion planning problem to a regular search for a shortest path in a graph, it can directly benefit from future advances in graph search algorithms or advances in the construction of more informative heuristics, which can be easily integrated into the developed framework.

5.6 Summary

This chapter has explained the construction of a search graph based on the multi-resolution state $\times$ time $\times$ goal lattice with hybrid dimensionality in order to reduce the motion planning problem to a regular search for a shortest path in a graph. For this purpose, conditions for the transition between regions with different resolution and dimensionality levels have been derived. The search graph has been extended to allow for combined planning via multiple waypoints.

Especially heuristic search algorithms are very well suited for efficiently finding shortest paths in the obtained graphs. One can choose from a wide range of established algorithms. In this thesis, the ARA* algorithm is used for the actual implementation and therefore it has been presented in detail in this chapter. Like any other heuristic graph search algorithm, ARA* requires the definition of a suitable cost function and informative heuristics. Since especially the latter is a prerequisite for the successful real-world application of the algorithm, multiple-waypoint heuristics that are both fast to compute and obstacle-aware have been developed. A discussion of the algorithm's complexity and the implications for the expected performance of the algorithm concluded the chapter.

Chapter 6

Results and Analysis

The previous chapters have established the theory for the proposed motion planning algorithm. This chapter provides information on the actual implementation of the motion planning framework and the results obtained both in simulation and for real-world applications. The quality of the results is evaluated and compared with other state-of-the-art planning algorithms.

6.1 Implementation

To ensure maximum portability, the proposed planning approach has been implemented in the form of a platform-independent C++ library. The availability of an efficient implementation for finding shortest paths in a graph is an essential requirement for several components of the proposed motion planning framework, since it is utilized for the decomposition of motion primitives (see Section 3.5.5), the computation of 2D obstacle-aware heuristics (see Section 5.4.2), and, of course, the main motion planning itself (see Section 5.2). All these algorithms maintain an internal priority queue to manage the list of states that may be expanded during the search (e.g., the *OPEN* list in Algorithm 2, p. 211). The most frequent operations on the priority queues are the insertion of new elements (INSERT), the extraction of the element with highest priority, i.e., lowest f-value, (EXTRACTMIN), increasing the priority of an element, i.e., decreasing the f-value, (DECREASEKEY), and checking whether an element is contained in the queue (FIND). In order to perform these operations efficiently, a specially tailored data structure based on a combination of Fibonacci heap and hash map [Cor09] has been developed in this thesis. It makes use of the Fibonacci heap implementation provided by the C++ library *Boost.Heap* [Ble11a]. It allows to perform the operations INSERT and DECREASEKEY with an amortized complexity of $O(1)$ and EXTRACTMIN with an amortized complexity of $O(\log n)$, where n is the number of elements in the heap [Ble11b]. The heap data structure is combined with a hash map whose implementation is taken from the C++ library *Boost.Unordered* [Jam08]. The hash map associates each state with its corresponding element in the heap and thus allows for an $O(1)$ implementation of the FIND operation. The presented priority queue implementation forms the basis for various graph search algorithms used in the proposed motion planning framework: The A* algorithm is used for the motion primitive decomposition, a Dijkstra-like algorithm is used for the computation of the 2D obstacle-aware heuristics, and Anytime Repairing A* (ARA*) is used for the actual motion planning through the state $\times$ time $\times$ goal lattice.

The developed motion planning library can be easily employed for various applications. In the context of this thesis, a MEX[1] wrapper has been developed, which enables the use of the proposed planning algorithms from within MATLAB. This allow comprehensive offline simulations of planning scenarios (see Section 6.3) since the algorithm's internals can be easily accessed for analysis and visualization purposes.

1. Mechanism to extend MATLAB with custom C, C++, or Fortran functions.

The second use case of the motion planning library is its integration into the Robot Operating System (ROS) [ROS16]. For this purpose, a thin wrapper with ROS-specific functionality has been developed around the C++ motion planning library. ROS is the de facto standard for middlewares in robotics. The framework relies on the concept of modularization and provides an efficient infrastructure for the communication between individual software components [Qui09]. For example, the developed motion planning module uses the infrastructure to receive data from the perception module and forward the planned motion to the trajectory controller module. The modular approach of ROS allows a straightforward application of the proposed motion planning algorithm to a real-world demonstrator (see Section 6.4).

6.2 Construction of Lattices and Motion Primitive Sets

As with all search-based planning algorithms, there exists a tradeoff between search speed and quantization fidelity. A coarse quantization enables rapid exploration of the search space, whereas a fine quantization allows for a more accurate representation of feasible robot motions. This is especially important in order not to miss any existing solution since planning in state lattices is only resolution-complete. The proposed multi-resolution concept tries to get the best out of those two worlds. Nonetheless, a concrete discretization scheme has to be chosen for the actual application. Good guidance is provided by the physical dimensions of the robotic platform: The mobile robot (see Section 3.4.1), which has been used for testing purposes both in simulation and experiments, has a size of $2.2\,\mathrm{m} \times 1.3\,\mathrm{m}$. The robot should be able to travel through narrow passages with about $0.5\,\mathrm{m}$ remaining space to each side. Furthermore, the map of the environment is assumed to have a resolution of $0.1\,\mathrm{m} \times 0.1\,\mathrm{m}$ per cell.

Table 6.1: Parameters for sampling of motion primitive sets used in the evaluation in Sections 6.3 to 6.5.

Parameter	Description	Value	Unit
State lattice configuration			
d	Dimensionality levels	$\{0, 1, 2\}$	
r	Resolution levels	$\{0, 1\}$	
High-resolution motion primitive set			
δ_{xy}^0	Position increment	0.2	m
$\lvert\Theta^0\rvert$	Number of heading values	32	
V^0	Set of velocities	$\{0, 1, 2\}$	m/s
δ_t^0	Time increment	0.25	s
$\Delta t_{\mathrm{m,max}}^0$	Max. motion primitive duration	1.5	s
Low-resolution motion primitive set			
δ_{xy}^1	Position increment	0.6	m
$\lvert\Theta^1\rvert$	Number of heading values	16	
V^1	Set of velocities	$\{0, 1, 2\}$	m/s
δ_t^1	Time increment	0.25	s
$\Delta t_{\mathrm{m,max}}^1$	Max. motion primitive duration	2.0	s
Resolution-independent			
N_{total}	Number of samples per bunch	1×10^8	
N_{expl}	Exploration threshold	5×10^7	
$e_{\mathrm{q,max}}$	Max. quantization error	0.2	
α	Weighting factor in (3.50)	0.002	
ϵ_{d}	Admissible cost increase for decomp.	1.02	
Robot model			
κ	Kinematical constant in (3.20)	1.47	m^{-1}
a	Admissible accelerations	$[-5, 5]$	$\mathrm{m/s^2}$
β	Admissible steering angles	$[-0.35, 0.35]$	rad

Considering these constraints, a position discretization $\delta_{xy}^0 = 0.2\,\text{m}$ is a sensible choice for the high-resolution planning portions ($r = 0$). To speed up planning in large free regions of the map, an additional resolution level ($r = r_{\text{min}} = 1$) is employed with $\delta_{xy}^1 = 0.6\,\text{m}$. For the discretization of the robot's heading, 16 distinct orientations for the low-resolution planning and 32 orientations for the high-resolution planning proved to be a reasonable choice (see [Pet13b]). The heading values are defined using the discretization scheme (3.27), which optimizes for smooth motion along the specified orientations. The complete list of all sampling parameters for the construction of the motion primitive sets is provided in Table 6.1 for reference. For the definition of the multi-resolution state × time lattice with hybrid dimensionality a total of three distinct dimensionalities (i.e., $d_{\text{min}} = 2$) are used in correspondence with the projection scheme (3.56) to (3.58). The construction of the motion primitive sets is performed according to the method outlined in Sections 3.5 to 3.7. Table 6.2 provides some statistics of the obtained motion primitive sets, which constitute the basis for the evaluation of the proposed motion planning concept in the remainder of this chapter. The table shows the total number of motion primitives per motion primitive set ("count"), the average number of motion primitives per bunch ("avg. bunch size"), and the average length of a motion primitive ("avg. length") both before and after the decomposition process.

6.3 Evaluation: Simulation Results

This section focuses on the evaluation of the proposed planning algorithm with respect to its functionality and capabilities. In order to show the correct functioning and effectiveness of the planning approach, extensive simulations have been carried out. For this purpose, the developed planning software has been integrated into MATLAB (see Section 6.1). On the

Table 6.2: Motion primitive set characteristics.

	Before decomposition			After decomposition		
$\mathcal{M}^{d,r}$	Count	Avg. bunch size	Avg. length	Count	Avg. bunch size	Avg. length
$\mathcal{M}^{0,0}$	4904	51.1	1.05 m	4556	47.5	1.04 m
$\mathcal{M}^{1,0}$	4136	43.1	1.06 m	2876	30.0	1.03 m
$\mathcal{M}^{2,0}$	1176	36.8	1.21 m	624	19.5	1.13 m
$\mathcal{M}^{0,1}$	596	12.4	1.25 m	472	9.8	1.19 m
$\mathcal{M}^{1,1}$	368	7.7	1.40 m	316	6.6	1.39 m
$\mathcal{M}^{2,1}$	96	6.0	1.62 m	96	6.0	1.62 m

basis of a previously recorded map of the environment, arbitrary scenarios consisting of the robot's current state $\mathbf{s}_s$, N waypoints g_i, $i = 1, \ldots, N$, and possibly some dynamic obstacles can be specified. The map of the environment and the robot's location therein are assumed to be perfectly known. This allows an undistorted evaluation of the motion planning results without potential side effects of other involved algorithms. For the very same reason, no additional trajectory smoothing or any post-processing is applied. The following figures directly show the result of a single planning query. The execution of the planned motion is not included in the simulation since this would involve further algorithms like a trajectory tracking controller, which might distort the actual planning results. The complete algorithmic chain for autonomous driving is described in Section 6.4 together with some real-world results of the proposed planning framework.

The holistic motion planning approach that has been developed in this thesis is capable of simultaneously planning in a hybrid-dimensional search space with multiple resolutions and along multiple waypoints while considering dynamic obstacles. Although this unified planning scheme is the key feature of the proposed approach, each of the following scenarios highlights one specific aspect for the sake of clarity. Nonetheless, all components of the planning algorithm are always active during the planning for each scenario.

Unless stated otherwise, the planning uses the temporal thresholds $\tau_0 = 3\,\mathrm{s}$ and $\tau_1 = 6\,\mathrm{s}$, i.e., dynamic obstacles are considered during the time-parametrized planning of the first three seconds. The next three seconds still consider the vehicle dynamics while after six seconds planning changes to mere kinematical planning. The respective dimensionality of the trajectory is visualized by graduated shadings of the robot's footprint (see Figure 6.2). The costs of the planned trajectories are computed according to (5.21) with $\eta_\mathrm{t} = 0.1$ and $\eta_\mathrm{r} = 10$. Besides the strong preference to avoid any risk, the planning focuses on a short trajectory. The duration of the trajectory is of lower priority because the accuracy of the estimated trajectory duration is limited due to the projection process of the motion primitives. Additionally, driving backwards is penalized by the factor $\eta_\mathrm{b} = 1.5$. The chosen weights for the individual components of the cost function serve as an example and may, of course, be changed according to the user's preference or the task's demands. This section focuses on the universal applicability of the proposed planning concept and therefore does not try to tune the parameters for any specific scenario.

The occupancy grid map of the environment is processed according to the method explained in Section 4.1, which defines risk zones for collision with static obstacles. Dynamic obstacles are represented by the time slices that have been introduced in Section 4.3.2. Finally, the high-resolution planning regions are determined using the method from Section 5.1.4. All simulations are performed on a desktop computer with an Intel Xeon E5-2687W CPU. Computing the time slices takes about 5 ms. The computational effort for the risk zone creation and determination of high-resolution planning regions depends on the map size (cell count) and is visualized in Figure 6.1. These computations need to be performed only when a new map of the environment is available. Thus, the computation

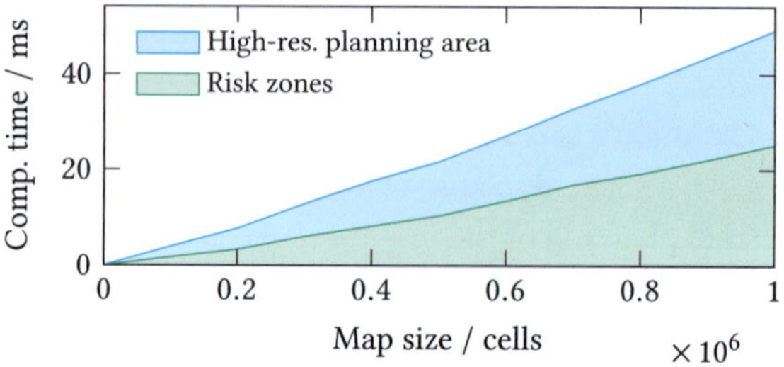

Figure 6.1: Computation time for map processing.

time listed for the following scenarios denotes the pure planning time without the computation time for map processing.

6.3.1 Hybrid Dimensionality

Figure 6.2 shows the result of high-resolution planning ($r = 0$), i.e., using only $\mathcal{M}^{0,0}$, $\mathcal{M}^{1,0}$, and $\mathcal{M}^{2,0}$ in an unstructured environment. This scenario serves as an example to explain the hybrid-dimensional planning scheme and the anytime character of the ARA* planning (cf. Section 5.2.2). All simulations use an initial heuristic inflation factor $\epsilon = 2$ for the ARA* search. The corresponding result of the first ARA* iteration is shown in blue in Figure 6.2. This initial trajectory is generally suboptimal; however, it is still a perfectly valid solution, i.e., it is guaranteed to avoid any collision with static or dynamic obstacles. This is exactly why anytime search algorithms are superior to classical planning algorithms in this context: They are capable of quickly generating avoidance maneuvers in critical situations. After having found the initial solution, the heuristic inflation factor ϵ is successively decreased by 0.05 in each ARA* iteration until the optimal[2]

2. Unless stated otherwise, the term *optimal* denotes the resolution-optimal solution with respect to the possible motions defined by the motion primitive sets and the utilized cost function. For $\epsilon = 1$, ARA* is guaranteed to find the shortest path in the search graph (see Section 5.2.2).

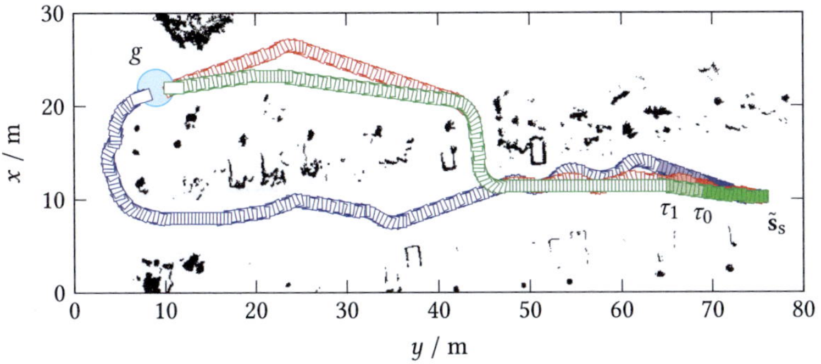

Figure 6.2: ARA* planning in a high-resolution lattice. ARA* iterations: ■■□ $\epsilon = 2$, ■■□ $\epsilon = 1.1$, ■■□ $\epsilon = 1$. Dimensionality of hybrid solution: ■■■ $\tilde{\mathbf{s}}_i^{0,0} \in L^{0,0} = X^0 \times Y^0 \times \Theta^0 \times V^0 \times T^0$, ■■■ $\tilde{\mathbf{s}}_i^{1,0} \in L^{1,0} = X^0 \times Y^0 \times \Theta^0 \times V^0$, □□□ $\tilde{\mathbf{s}}_i^{2,0} \in L^{2,0} = X^0 \times Y^0 \times \Theta^0$.

solution is eventually found for $\epsilon = 1$ (depicted in green). Additionally, the intermediate solution for $\epsilon = 1.1$ is shown (red).

Besides the anytime planning scheme, Figure 6.2 illustrates the hybrid nature of the solutions: The dark tiles at the beginning indicate a full-dimensional trajectory (including the temporal dimension), which is the result of time-parametrized planning in $L^{0,0}$ until the temporal threshold τ_0. This initial segment of the solution corresponds to the planning segment that is able to consider dynamic obstacles (for examples see Section 6.3.3). The next segment (medium-shaded tiles) corresponds to planning in $L^{1,0}$, which still respects the dynamics of the vehicle. At τ_1 the planning transitions to mere path planning in $L^{2,0}$ (white tiles) considering only the vehicle's kinematics. One can see that the segment from τ_0 to τ_1 is considerably shorter than the first segment to τ_0. This is due to the projection process (3.57) in which the temporal dimension is removed and thus a motion primitive might be superseded by a motion primitive with longer duration but shorter length due to the scoring with $\eta_t < 1$.

Figures 6.3 and 6.4 show the expanded nodes for the respective ARA* iterations of the scenario depicted in Figure 6.2. More precisely, Figure 6.3 shows the accumulated costs of each expanded state. For visualization purposes, all costs are projected onto the xy-plane by

$$\underline{g}(\tilde{x}, \tilde{y}) := \min_{\tilde{\theta}, \tilde{v}, \tilde{t}} g(\tilde{\mathrm{s}}^{d,r}), \tag{6.1}$$

i.e., the diagram shows the minimum cost for each position under all possible headings, velocities and times (if applicable). Figure 6.4, on the other hand, shows the heuristic costs $h(\tilde{\mathrm{s}}) = h(\tilde{x}, \tilde{y})$ according to (5.26). For the unbiased assessment of the planning result, the planning uses a simple Euclidean heuristic. The effect of using the improved 2D obstacle-aware heuristic is discussed at the end of this section.

Figure 6.5 shows some characteristics of the ARA* planning process. The most interesting parameter is the cumulative computation time, which is the sum of the computation time for all ARA* iterations up to the currently considered ϵ. The initial solution for $\epsilon = 2$ is available after 85 ms. A total of 4.6 s pass until the optimal solution is found for $\epsilon = 1$. The latter contradicts the requirement for a planning frequency of approximately 10 Hz and shows that planning in a high-resolution lattice only is not practical. This is where multi-resolution planning comes into action (see next section). Furthermore, the suboptimality bound ϵ', defined according to (5.18), is shown in Figure 6.5. The very first ARA* iteration for $\epsilon = 2$ is able to find a solution whose cost is guaranteed to exceed the optimal cost for $\epsilon = 1$ by at most the factor $\epsilon' = 1.33$. The accumulated cost to the goal, $g(\tilde{\mathrm{s}}_g)$, decreases from 91.02 for $\epsilon = 2$ to 74.38 for $\epsilon = 1$ during the ARA* iterations. The observed data in each ARA* iteration is visualized by points in the plot; the connecting lines are only shown for visualization purposes.

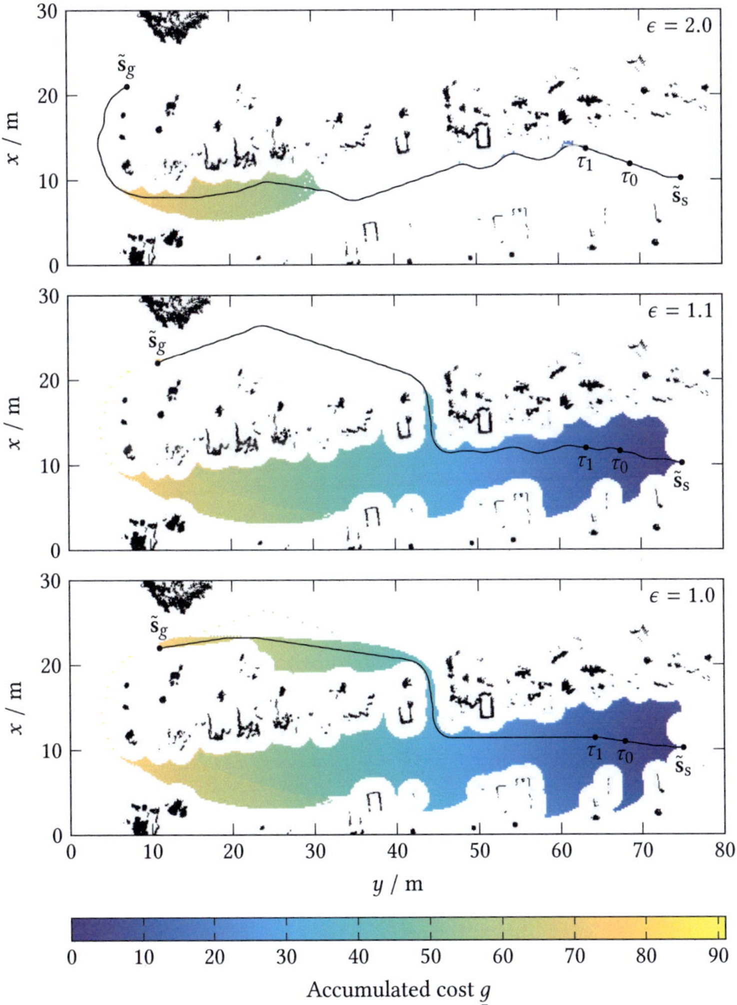

Figure 6.3: Accumulated cost g of expanded nodes for the ARA* iterations of the high-resolution scenario from Figure 6.2.

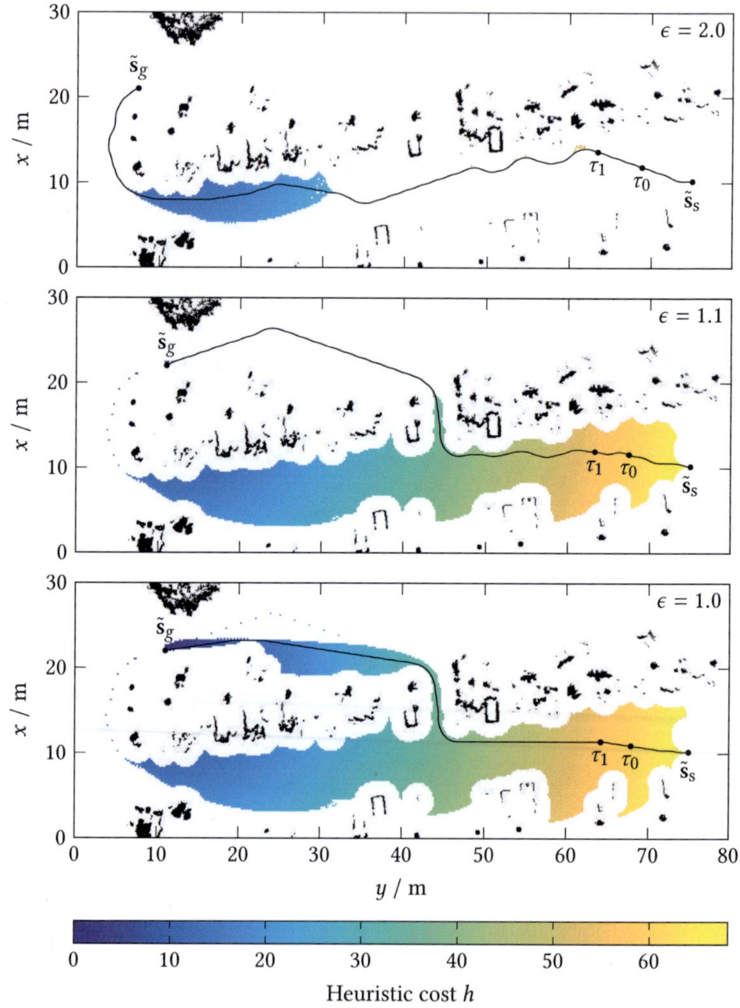

Figure 6.4: Heuristic cost h of expanded nodes for the ARA* iterations of the high-resolution scenario from Figure 6.2.

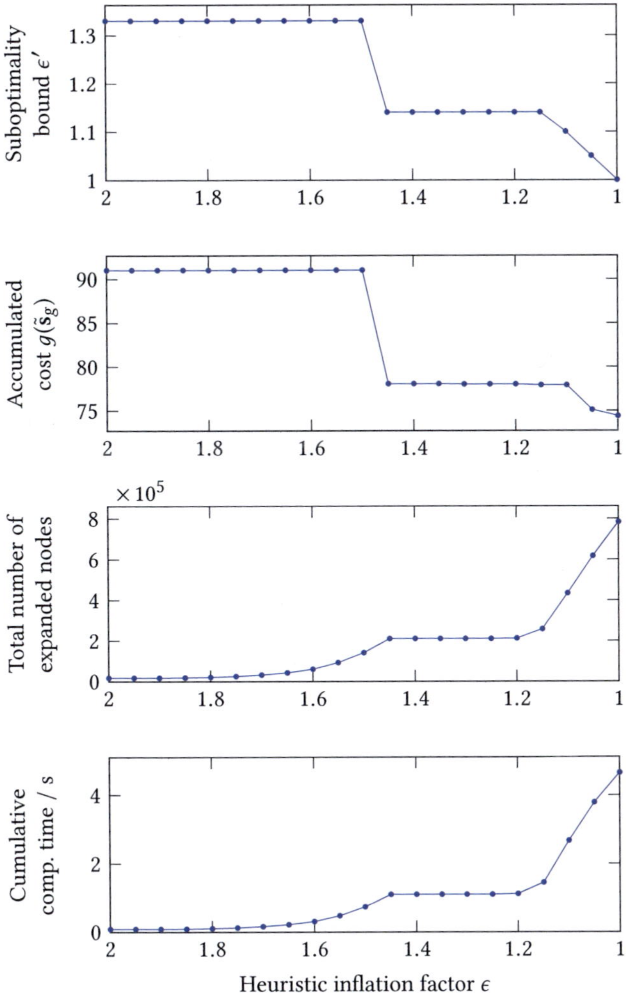

Figure 6.5: Characteristics of ARA* iterations for the high-resolution scenario from Figure 6.2.

In the high-resolution planning scenario depicted in Figures 6.2 to 6.4, the choice of the planning dimensionality was made on the basis of the temporal thresholds $\tau_0 = 3\,\mathrm{s}$ and $\tau_1 = 6\,\mathrm{s}$. Figure 6.6 shows how the computation time depends on these thresholds. As one would expect, the computation time increases for greater thresholds since this results in planning with high dimensionality for a longer period. It can also be seen that the computation time for $\epsilon \leq 1.4$ is nearly constant. This is caused by the strong overestimation of the heuristics, which forces the graph to connect to the goal as fast as possible.

The required planning time can be reduced by using the improved obstacle-aware 2D heuristics from Section 5.4.2. The benefit of using the obstacle-aware 2D heuristic with respect to its computation threshold $\bar{\mathfrak{d}}_{\mathrm{obst}}$ (see p. 141) is visualized in Figure 6.7. It can be seen that for very small computation thresholds, the total planning actually increases. This is caused by the heuristic inflation factor $\epsilon > 1$ during most of the ARA* planning, which deliberately violates the admissibility of the heuristics. From $\bar{\mathfrak{d}}_{\mathrm{obst}} = 27\,\mathrm{m}$ on, the planning time decreases steadily in this scenario (except for small oscillations). If the computation of the obstacle-aware 2D heuristic extends all the way from the goal to the start, the planning time can be reduced from 4.6 s to 1.4 s, which is a reduction by 70 %.

6.3.2 Multiple Resolutions

In order to demonstrate the advantages of the multi-resolution planning approach, this section presents the planning results for the very same scenario laid out in the previous section but now with utilization of all motion primitives from Table 6.2. Both task- and environment-based criteria are applied for the determination of high-resolution planning regions: High-resolution planning takes place within a radius of 2 m around the start and goal, and, additionally, further environment-specific high-resolution

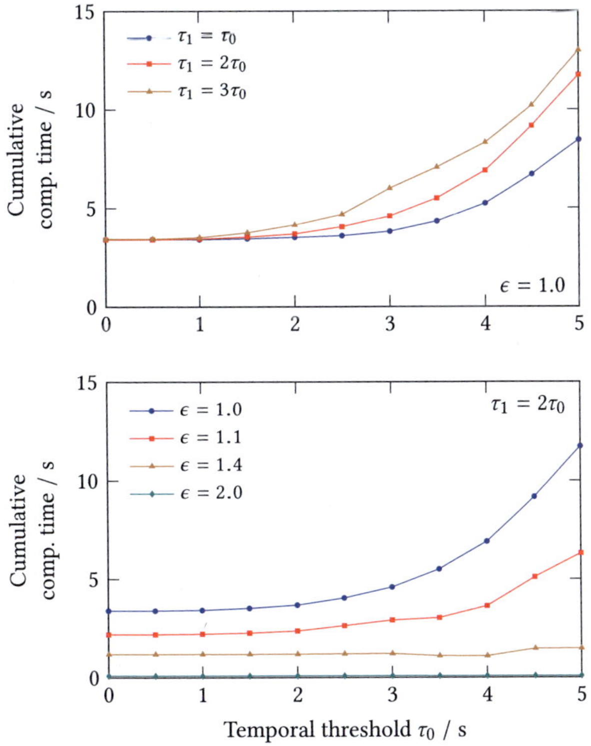

Figure 6.6: Influence of temporal thresholds on computation time in the high-resolution scenario from Figure 6.2.

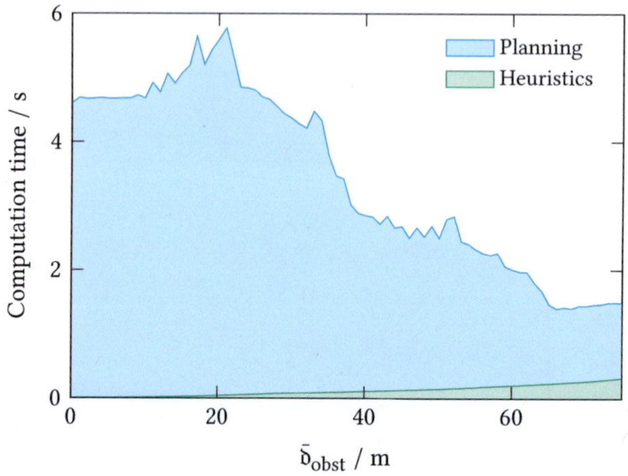

Figure 6.7: Influence of distance threshold $\bar{\delta}_{obst}$ on computation time.

planning regions are computed on the basis of the method described in Section 5.1.4.

Figure 6.8 illustrates the result of the multi-resolution planning so that it can be easily compared to the high-resolution planning in Figure 6.2. Once again, the (intermediate) ARA* solutions for $\epsilon = 2$, $\epsilon = 1.1$, and $\epsilon = 1$ are shown. It can be seen that the optimal multi-resolution trajectory does not fully reach the quality of the optimal high-resolution trajectory. This is primarily due to the lesser number of discrete headings in the low-resolution planning regions. Nonetheless, the multi-resolution trajectory is a perfectly usable solution, whose cost increased only slightly from 74.38 for the high-resolution planning to 76.32 for the multi-resolution planning, which amounts to 2.6 % (see Figure 6.13). Moreover, only the high-resolution part of the trajectory is actually executed by the robot in each cycle due to the continuous replanning approach.

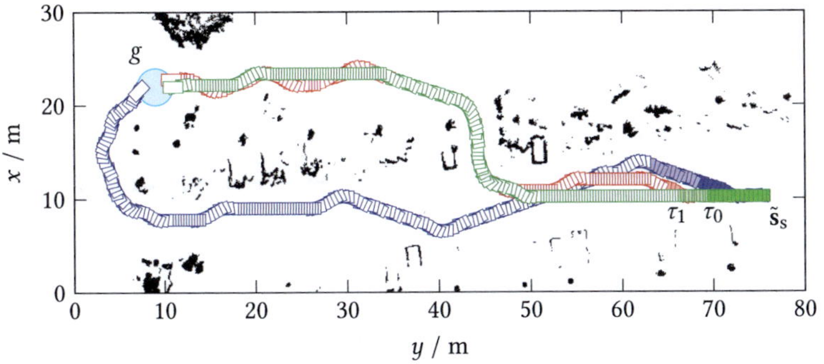

Figure 6.8: ARA* planning in a multi-resolution lattice. Computation time 67 ms. ARA* iterations: ■■□ $\epsilon = 2$, ■■□ $\epsilon = 1.1$, ■■□ $\epsilon = 1$. Dimensionality of hybrid solution: ■■■ $\tilde{s}_i^{0,r} \in L^{0,r} = X^r \times Y^r \times \Theta^r \times V^r \times T^r$, ■■■ $\tilde{s}_i^{1,r} \in L^{1,r} = X^r \times Y^r \times \Theta^r \times V^r$, □□□ $\tilde{s}_i^{2,r} \in L^{2,r} = X^r \times Y^r \times \Theta^r$.

Figures 6.9 and 6.10 show the expanded states analogously to Figures 6.3 and 6.4 for three ARA* iterations. Again, the projection (6.1) has been performed for the accumulated costs in Figure 6.9 for visualization purposes. Figure 6.10 shows the heuristic costs, which do not depend on the resolution and are thus identical to the heuristic costs in 6.4. The regions with different planning resolutions are clearly distinguishable in the figures: While the largest part of the search space is processed by low-resolution planning, high-resolution planning is employed near the start and goal and in the narrow passage in the center of the map.

Figure 6.11 shows some characteristics of the multi-resolution planning for comparison with the corresponding high-resolution planning characteristics in Figure 6.5. The multi-resolution planning is able to find an initial solution after only 1 ms. After a total of 67 ms, the optimal solution for $\epsilon = 1$ is found. Thus, for the given scenario, the multi-resolution planning could reduce the total planning time to 1.5 % of the high-resolution planning time.

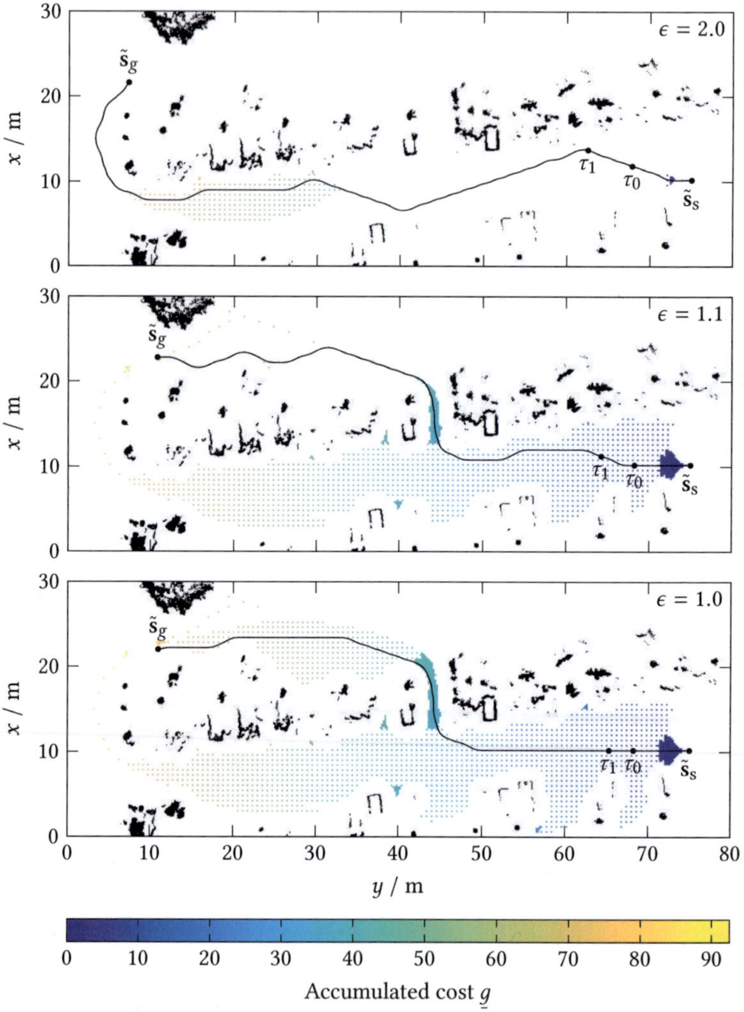

Figure 6.9: Accumulated cost $\underline{g}$ of expanded nodes for the ARA* iterations of the multi-resolution scenario from Figure 6.8.

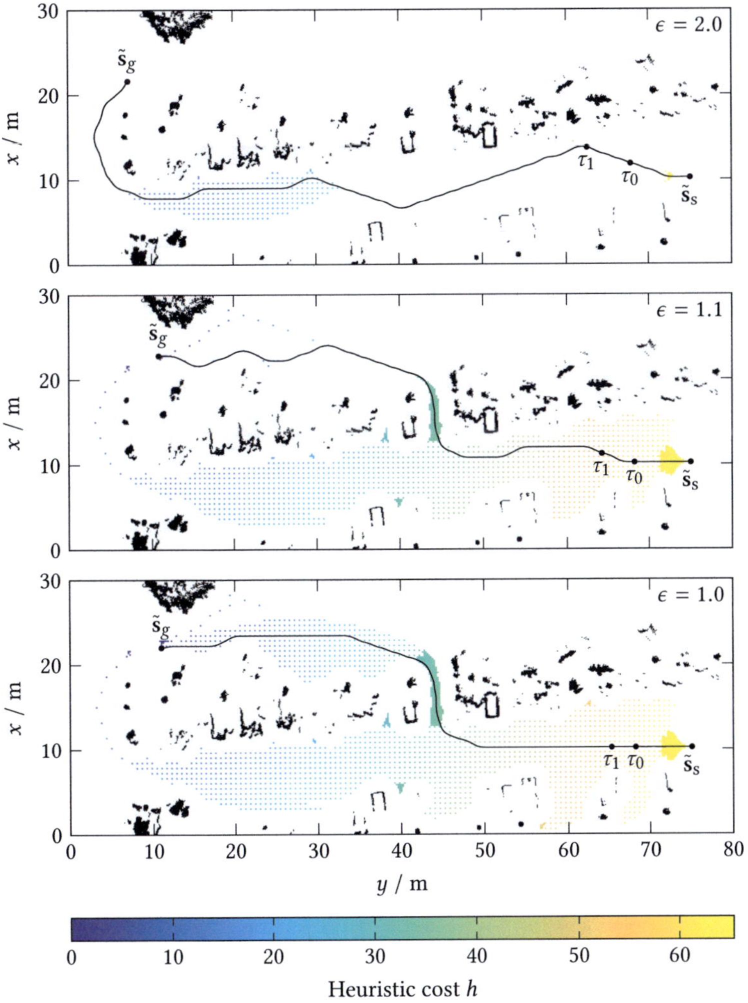

Figure 6.10: Heuristic cost h of expanded nodes for the ARA* iterations of the multi-resolution scenario from Figure 6.8.

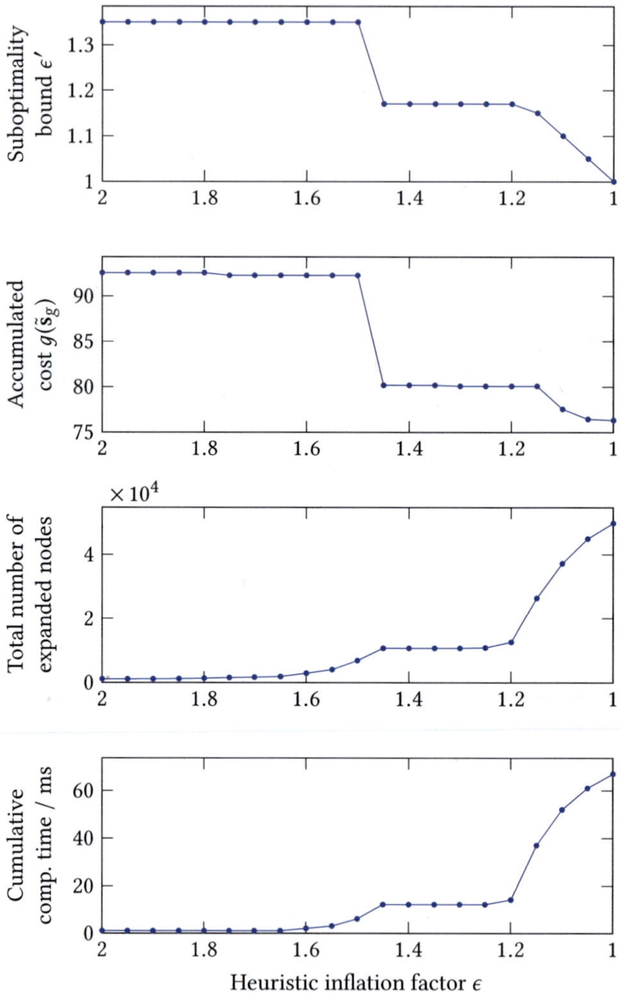

Figure 6.11: Characteristics of ARA* iterations for the multi-resolution scenario from Figure 6.8.

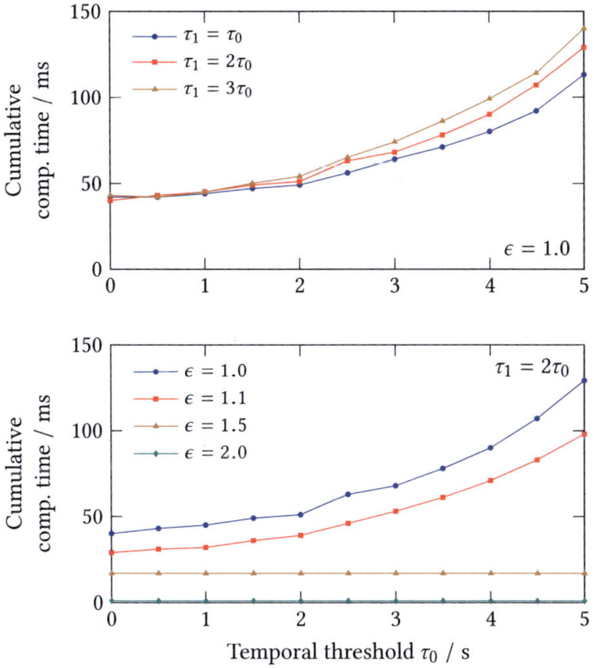

Figure 6.12: Influence of temporal thresholds on computation time in the multi-resolution scenario from Figure 6.8.

As expected, computation time increases with greater temporal thresholds τ_0 and τ_1 also in case of multi-resolution planning (see Figure 6.12). For this particular scenario, the intended planning frequency of 10 Hz can be achieved for thresholds below $\tau_1 = 2\tau_0 = 8$ s.

The cost of a solution decreases with increasing size of the high-resolution planning regions. This property is guaranteed by condition (3.66). Figure 6.13 shows the influence of this high-resolution planning region extent on solution cost and computation time. The extent is specified by the

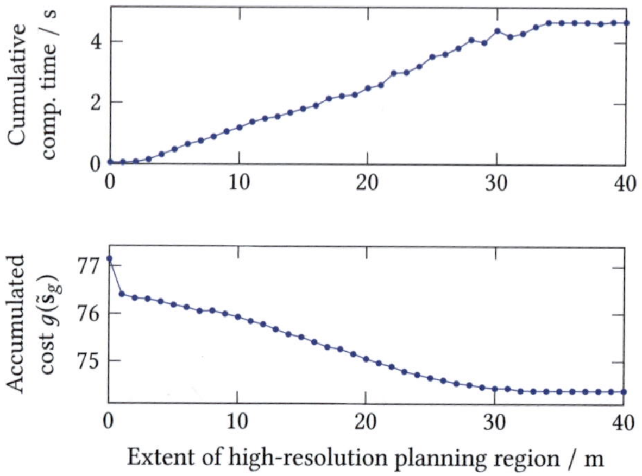

Figure 6.13: Influence of high-resolution planning area extent in the multi-resolution scenario from Figure 6.8 for $\epsilon = 1.0$.

radius of the high-resolution planning area at the start and goal (task-based criterion). It can be seen that from 35 m on the computation time does not increase any more and the solution cost remains the same. This is because for this radius the start and goal high-resolution planning regions meet and thus the complete planning takes place in the high-resolution lattice. With this, the planning becomes identical to the planning in Section 6.3.1.

An additional multi-resolution planning scenario is depicted in Figure 6.14: The task is to plan a motion that moves the robot into the narrow parking space with a fixed end orientation (namely rear-facing). This requires, of course, a motion primitive set that allows the robot to drive forward and backward, which is why this scenario uses a motion primitive set that was sampled with $V^r = \{ -2\,\mathrm{m/s}, 0\,\mathrm{m/s}, 2\,\mathrm{m/s} \}$; the other parameters from Table 6.1 remained unchanged. The purpose of this scenario is to demonstrate the necessity of multi-resolution planning: The lattice points of the low-resolution lattice are too close to the boundary of the

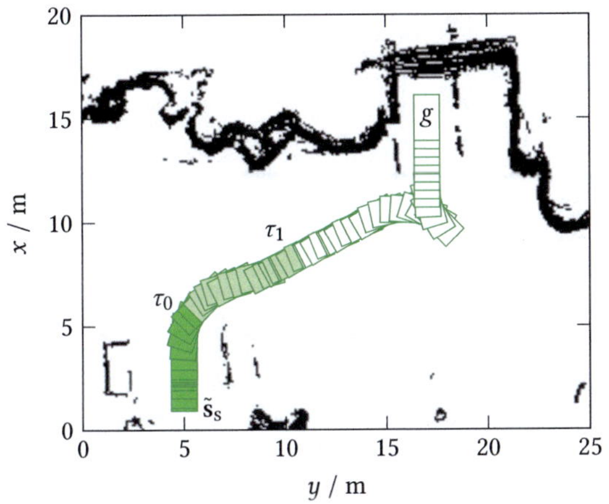

Figure 6.14: Multi-resolution planning of a reverse parking maneuver. Only the optimal solution for $\epsilon = 1$ is shown.

parking space, which prevents low-resolution-only planning from finding a solution. Only the high-resolution planning near the goal enables the successful planning of the parking maneuver. The multi-resolution planning finds a solution (see Figure 6.14) in 85 ms while the full high-resolution planning would have taken 1.4 s.

6.3.3 Dynamic Obstacles

This section provides some results for motion planning in the presence of dynamic obstacles. Figure 6.15 depicts a scenario in which the robot is located next to a dynamic obstacle, which is moving left with a velocity of 2 m/s. The scenario clearly shows why a holistic approach, which combines global path planning and local obstacle avoidance by time-parametrized motion planning techniques, is superior to separate planning and obstacle avoidance. Figure 6.15a shows the result of conventional planning, which only considers the obstacle's current position at $t = 0$. Relying on pure path planning would inevitably lead to a collision. Therefore, a subsequent algorithm is necessary in order to locally modify the planned path and thus avoid the collision. This local path or trajectory modification is often performed in a continuous fashion, which means that it cannot skip over static obstacles (see the small static obstacle (black) directly below the dynamic obstacle in Figure 6.15a). Instead, a local obstacle avoidance strategy might be to simply wait while the obstacle has passed. By contrast, Figure 6.15b shows the result of combined planning according to the hybrid-dimensional multi-resolution planning scheme. The planning incorporates the prediction of the dynamic obstacle from $t = 0$ to $t = \tau_0 = 3$ s. The resulting plan is a smooth avoidance maneuver which allows the robot to continue its journey without any risk for a collision with the dynamic obstacle. The dynamic obstacle is assumed to have circular shape and is modeled according to Section 4.3.1. The red uncertainty ellipses show the 90 % confidence intervals of the predicted obstacle location. The proposed planning concept allows the computation of safe local avoidance maneuvers while still optimizing the overall global plan.

A further example for planning in the presence of dynamic obstacles is shown in Figure 6.16. In this scenario, the obstacle is modeled using the concept of time slices devised in Section 4.3.2. This allows the correct consideration of the actual shape of the obstacle, which is often rectangular (other vehicles) instead of circular. It can be seen that during the full-dimensional planning phase the robot plans an avoidance maneuver and then waits until the obstacle has passed. The initial—slightly suboptimal—avoidance maneuver is available after only 26 ms. This is where the anytime nature of ARA* shows its strengths. In addition, it becomes clear from the planned motion that all components of the proposed planning approach are simultaneously active: Besides the utilization of variable di-

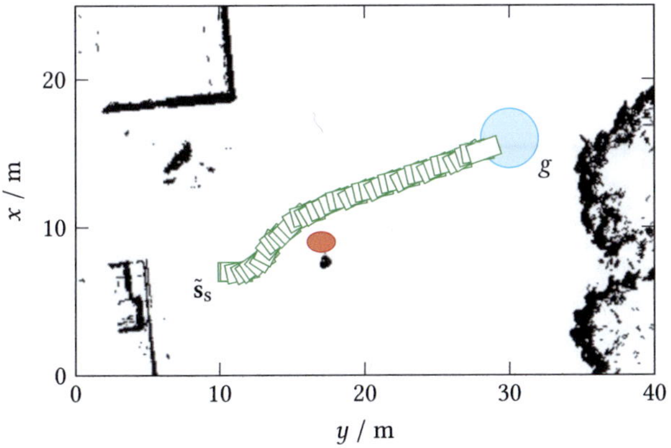

(a) Conventional path planning without obstacle prediction. Computation time less than 1 ms.

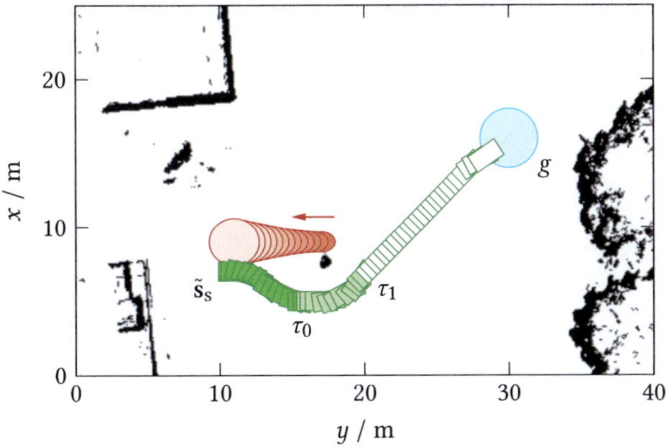

(b) Combined planning of local avoidance maneuver and global path. Computation time 43 ms.

Figure 6.15: Planning with and without dynamic obstacle prediction.

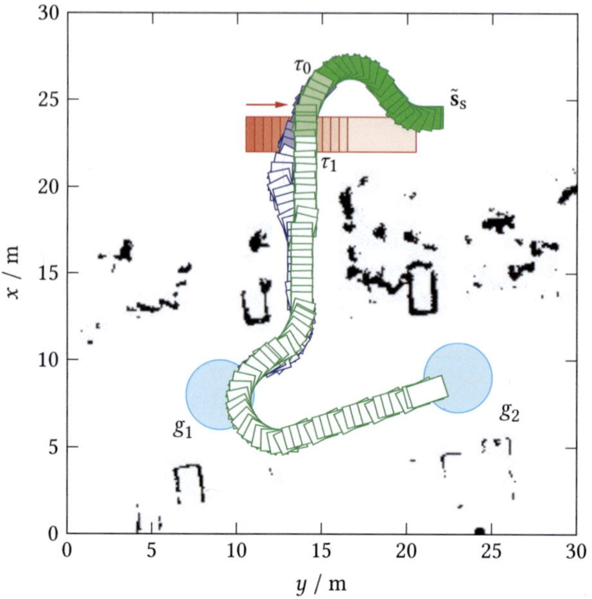

Figure 6.16: Dynamic obstacle avoidance based on time slices and planning for multiple waypoints simultaneously. An initial (suboptimal) avoidance maneuver (■■□, $\epsilon = 2$) is available after 26 ms. The optimal solution (■■□, $\epsilon = 1$) is found after a total of 75 ms.

mensionality and multiple resolutions, the planning is jointly performed for two waypoints. This is discussed in detail in Section 6.3.4.

The consideration of custom regions with increased collision risk is shown in Figure 6.17. The region extends from the bottom of the map to the top and divides it into two halves. Thus, the robot is required to traverse this risk region in order to reach the goal. The planned motion tries to minimize the collision risk by crossing the region with increased collision risk at the shortest way possible (i.e., perpendicularly).

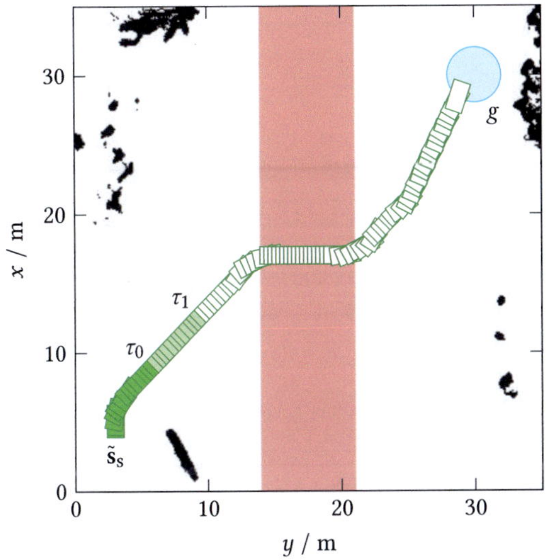

Figure 6.17: Planning through a region with increased collision risk (red zone). Computation time 68 ms.

6.3.4 Multiple Waypoints

The combined planning for multiple waypoints is a key feature of the proposed motion planning algorithm. This section highlights some advantages of multiple-waypoint planning in various scenarios.

Figure 6.18 shows the multi-resolution planning scenario from Figure 6.8, but now, additionally, a waypoint g_1 was added halfway between start $\tilde{s}_s$ and goal g_2. Since g_1 is located on the optimal path from $\tilde{s}_s$ to g_2, the solution for additionally planning via g_1 is identical to the solution from Figure 6.8. The required computation time slightly decreases from 67 ms for directly planning to g_2 to 61 ms for planning to g_2 via g_1. This is due to the fact that the intermediate waypoint g_1 directs the search toward the goal g_2 right from the beginning. The initial planning ($\epsilon = 2$, blue) is

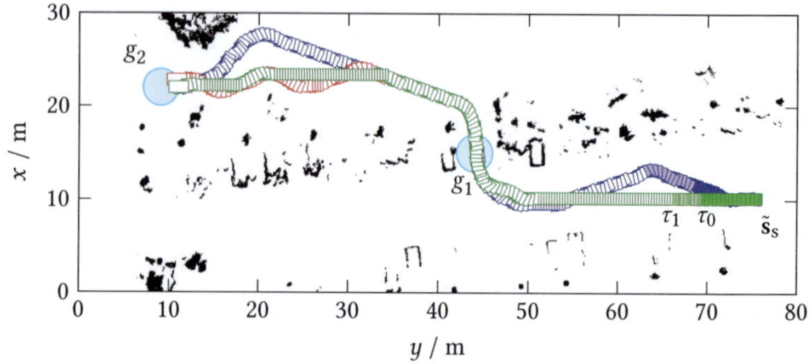

Figure 6.18: Multiple-waypoint planning in a multi-resolution lattice. Computation time 61 ms. ARA* iterations: ■■□ $\epsilon = 2$, ■■□ $\epsilon = 1.1$, ■■□ $\epsilon = 1$. Dimensionality of hybrid solution: ■■■ $\tilde{\mathbf{s}}_i^{0,r} \in L^{0,r} = X^r \times Y^r \times \Theta^r \times V^r \times T^r \times G$, ■■■ $\tilde{\mathbf{s}}_i^{1,r} \in L^{1,r} = X^r \times Y^r \times \Theta^r \times V^r \times G$, □□□ $\tilde{\mathbf{s}}_i^{2,r} \in L^{2,r} = X^r \times Y^r \times \Theta^r \times G$.

already close to the optimal solution (green). This scenario clearly shows that the extension of the state × time lattice to a state × time × goal lattice (i.e., the introduction of an additional goal dimension, see Section 5.1.5) does not necessarily increase the planning time, which only depends on the length and complexity of the solution.

A second example for multiple-waypoint planning is depicted in Figure 6.19. This basically corresponds to the motivating scenario in Figure 3.15. In Figure 6.19a, planning is performed to the next waypoint only, which results in the robot reaching the waypoint g_1 in a disadvantageous orientation: The robot needs to perform an unnecessary turning maneuver in order to continue its journey to waypoint g_2. Once again, the planning was performed using the same motion primitive sets as in the parking scenario in Figure 6.14. The discrete set of velocities $V^r = \{-2\,\mathrm{m/s}, 0\,\mathrm{m/s}, 2\,\mathrm{m/s}\}$ allows the robot to drive forward and backward. If the motion primitive sets did not allow for backward motion, the

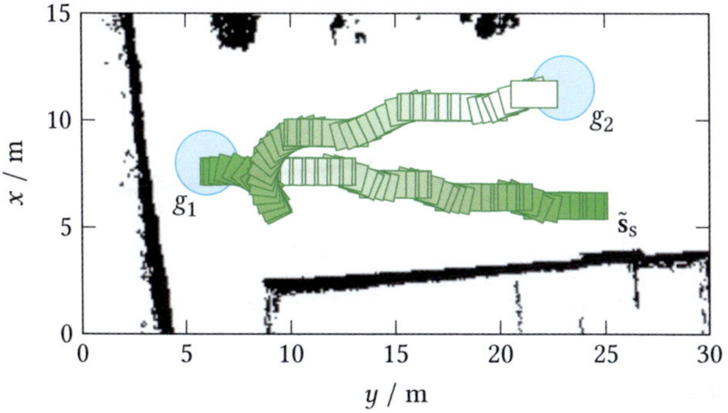

(a) Multi-waypoint scenario with separate planning. Computation time was less than 1 ms for the first segment and 51 ms for the second segment.

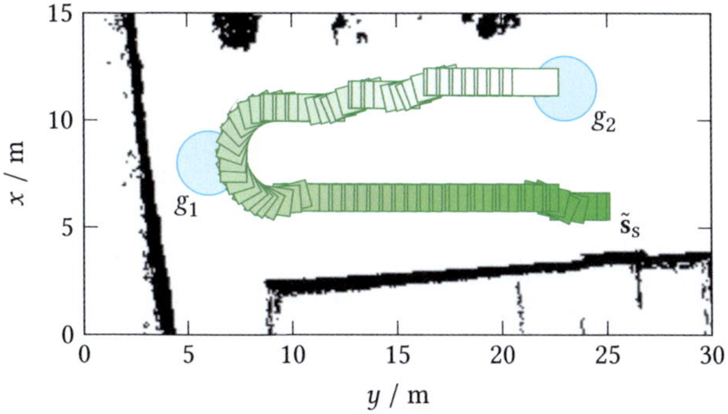

(b) Multi-waypoint scenario with combined planning. Total computation time was 95 ms.

Figure 6.19: Comparison of separate and combined multi-waypoint planning. The shading of the robot's footprint corresponds to the evolution of time along the planned segments to emphasize the two separate plans in (a).

robot's journey would even have ended at g_1 since the mobile robot could not free itself from this situation.

On the other hand, Figure 6.19b shows the combined planning for g_1 and g_2. One can see that the robot touches g_1 in an optimal way in order to continue its way to g_2. The shading of the robot's tiles in Figure 6.19 visualizes the time along the planned trajectory instead of its dimensionality so one can clearly see that the plan in Figure 6.19a consists of two separate segments whereas only one joint planning was performed in Figure 6.19b. In this scenario, the total planning time increased from 52 ms (Figure 6.19a) to 95 ms (Figure 6.19b) because the computation of the first part of the trajectory to g_1 becomes much more involved. The increase in the total planning time is considered reasonable as the unnecessary turning maneuver should typically be avoided. Like in the previous scenarios, all elements of hybrid-dimensional multi-resolution planning among dynamic obstacles have been active during the planning of the scenarios in this section.

6.4 Evaluation: Real-World Demonstrator

The results presented in Section 6.3 were obtained by simulations that were based on real recorded sensor data, which is why those results already exhibit high significance and are directly applicable to practice. Nonetheless, the proposed planning algorithm has additionally been implemented on several real robotic platforms in order to prove the practical suitability. One of them is the technology demonstrator *IOSB.amp Q2* (see Figure 6.20), which was developed at the Fraunhofer Institute of Optronics, System Technologies and Image Exploitation IOSB in Karlsruhe, Germany, during the course of this thesis.

This robotic vehicle is equipped with four-wheel steering and thus its kinematics correspond to the system model described in Section 3.4. With

Figure 6.20: Mobile all-terrain robot *IOSB.amp Q2*.

a hybrid drive, the robot has an operating range of 200 km and a permissible load capacity of 150 kg. The robotic platform is equipped with many sensors for environment perception and navigation. A 3D laser scanner generates 700 000 range values per second, which are used to build a map of the environment [Emt12; Emt14]. Besides additional line laser scanners, which have been installed for safety reasons, the vehicle is also equipped with an inertial measurement unit (IMU). The IMU comprises three accelerometers and three gyroscopes. These sensors are used together with wheel encoders and a GPS for localization purposes. The data of individual sensors is combined by means of sensor fusion to take advantage of the strengths of each sensor system [Emt10; Emt14]. The generated map of the environment and the estimated state of the robot serve as inputs for the motion planning module. A supervisory behavior control module specifies the list of waypoints and detects if a waypoint has been reached by the robot. The planned trajectory is finally tracked by a path controller for nonholonomic robots that is based on [DeL98]. The robot makes extensive use of the infrastructure provided by the Robot Operating System (ROS). The motion planning algorithm is implemented in terms of a ROS node which is a thin wrapper around the developed portable C++ library.

Figure 6.21 shows the recorded trajectory of the physical robot in a multi-waypoint scenario. Here, planning was performed for two waypoints at a time. In contrast to the illustrations in Section 6.3, which depict a single snapshot of the instantaneous planning result, Figure 6.21 shows the full traveled path, which is the result of alternating planning and control. The robot footprint is plotted every 0.5 s. It can be clearly seen that the robot's motion is very smooth and similar to what a human driver would achieve. At this point, it is worth mentioning again that the path controller directly operated on the output of the motion planning algorithm without employing any intermediate trajectory smoothing. The computation time of the motion planning algorithm during the run of this scenario is depicted in Figures 6.22 and 6.23. One can see that the computation time stays well below the intended 100 ms. The initial solution for 99 % of the computed plans was available after 25 ms and computing the optimal solution took 62 ms in total. Further aspects of the planning concept (like hybrid dimensionality, multiple resolutions, and multiple waypoints) are not discussed here since the observations on planning the motion of the physical robot are identical to the simulated results in Section 6.3 due to the identical software core.

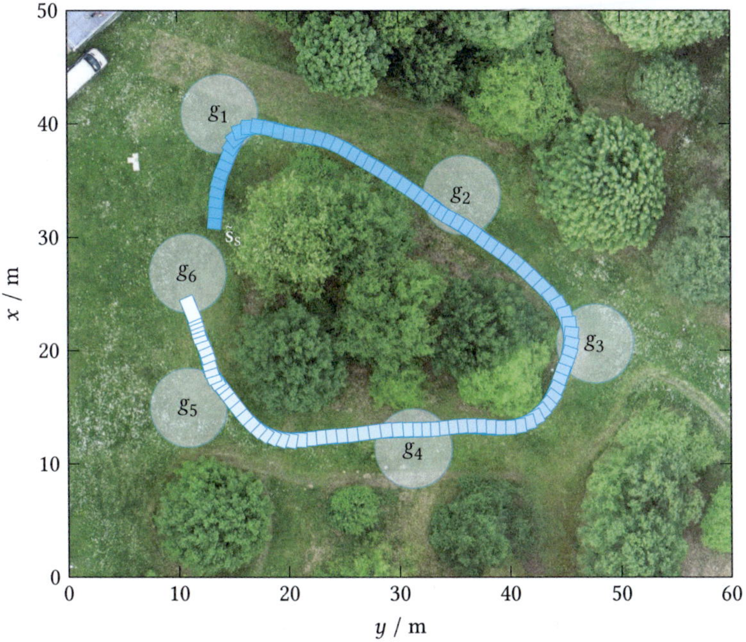

Figure 6.21: Planning result for a real-world scenario with six waypoints ($g_1, \ldots, g_6$). The result is visualized on top of a georeferenced orthomosaic obtained by means of photogrammetric methods. The shown robot footprints correspond to the actual motion and are thus the result of the whole data processing cycle. No intermediate trajectory smoothing was employed. The shading depicts the travel time from $t = 0$ s (cyan) to $t = 55$ s (white).

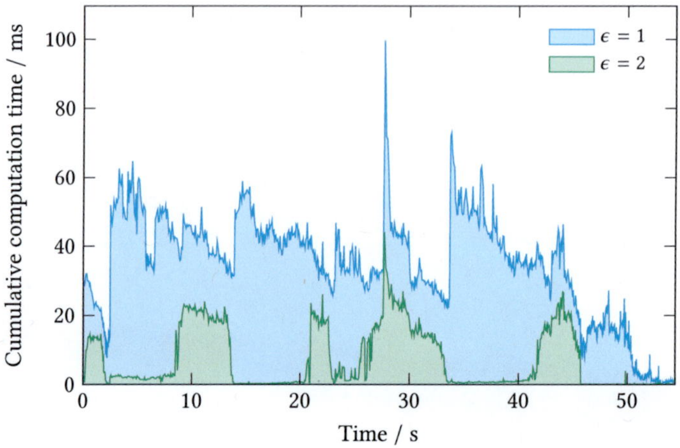

(a) Variation of cumulative computation time with travel time.

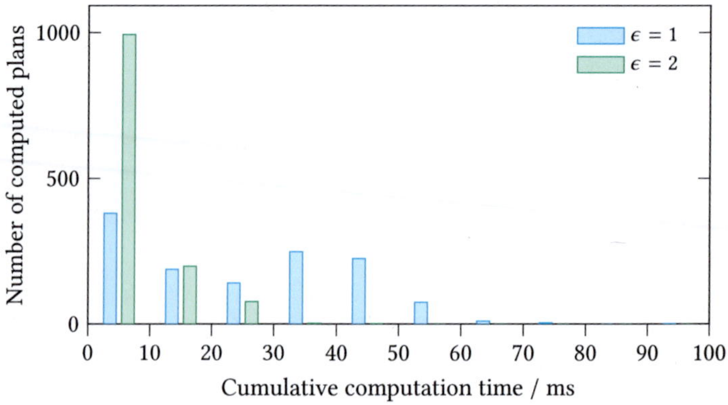

(b) Histogram of cumulative computation time.

Figure 6.22: Computation time for the scenario from Figure 6.21.

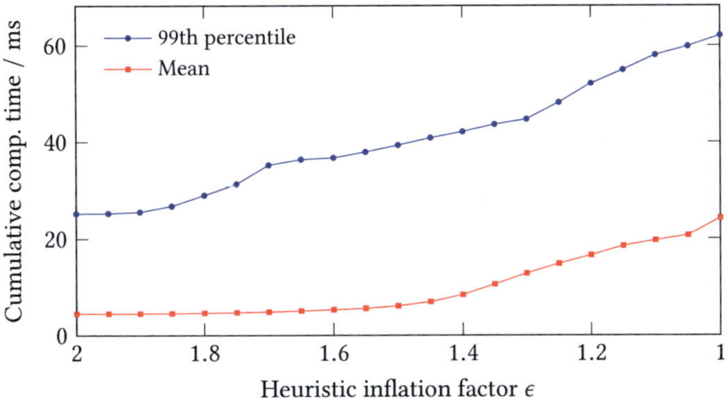

Figure 6.23: Variation of cumulative computation time with the iterative refinement of solutions, i.e., ARA* iterations with decreasing heuristic inflation factor ϵ, for the scenario from Figure 6.21.

6.5 Evaluation: Further Metrics

While the previous sections focused on the evaluation of the capabilities of the proposed planning algorithm, this section is concerned with the discussion of commonly considered evaluation metrics. The evaluation of motion planning algorithms for mobile robots is a challenging topic. The authors of [Bal00] acknowledge the fact that a meaningful performance measure strongly depends on the particular task of the robot and that verifiable statements require the consideration of the entire processing cycle (sensing → perception → reasoning → acting). But also the representation of the environment and the implementation of the algorithm itself (e.g., choice of programming language) may have a strong impact on performance. This is why evaluation is commonly conducted empirically in pertinent challenges like RoboCup, ELROB, or the DARPA challenges [Now10]. Nonetheless, the robotic community strives for quantitative assessment of single algorithms. For this purpose, various benchmark frameworks have been proposed for mobile robots in the past (e.g., [Bal00; Cal08; Muñ10]), which, however, did not gain broad acceptance due to the lack of standardized interfaces [Now10].

A promising approach is described in [Coh12]; however, it is strongly focused on manipulator motion planning and the applicability to mobile robots is somewhat limited (e.g., relatively simple environment representation). Nonetheless, the paper proposes some fundamental performance metrics which are also applicable to mobile robot motion planning: computation time, path length, minimum clearance, smoothness, and success rate. In addition, the authors of [Muñ10] suggest to also consider the occurrence of collisions and the robustness in narrow spaces. Further criteria are reaction time, number of parameters to tune and the precision for reaching the target [Now10]. The remainder of this section provides a detailed discussion of the mentioned performance measures with respect to the proposed planning algorithm.

Computation and reaction time

Computation time is the most implementation-dependent metric. The processing capabilities of the hardware, the employed programming language, as well as programming skills heavily influence the performance. The planning algorithm developed in this thesis has been implemented in C++ and has been evaluated on an Intel Xeon E5-2687W (see Section 6.1). It has already been mentioned in Section 6.4 that the software which runs on the real robot is identical to the software that was used for the simulations in Section 6.3 (besides a thin visualization layer). Furthermore, all simulations were based on real recorded sensor data. Thus, a thorough quantitative evaluation of the proposed planning algorithm with respect to the required computation time has already been conducted in Section 6.3, which explicitly stated the computation time for each scenario. For all hybrid-dimensional multi-resolution planning scenarios from that section, a solution was found in less than 100 ms. An analysis of the influence of the temporal thresholds on computation time is provided in Figure 6.12. All these computation times refer to the time which is required to compute the optimal solution for $\epsilon = 1$. An initial, possibly suboptimal, but still perfectly valid solution is available much faster. An example is given in Figure 6.16, where an initial avoidance maneuver ($\epsilon = 2$) is available after 26 ms, which is one-third of the total computation time for the optimal solution. The decrease in computation and thus reaction time in the presence of dynamic obstacles is due to the simultaneous employment of hybrid dimensionality and multiple resolutions (see Sections 6.3.1 and 6.3.2).

Path length

This metric, which assesses the optimality of the solution, originates from the field of probabilistic planning algorithms. The proposed planning algorithm produces, by design, plans that are always optimal with respect to the chosen discretization of the search space. Since the method computes hybrid solutions that are part trajectory and part path, considering only the path length is not enough. Instead, one needs to take the time necessary to execute the planned motion additionally into account. Although path length and execution time correlate, it is up to the user to specify any preferences by choosing appropriate penalties for the individual components of the cost function. This in turn depends on the particular task, which is why general figures are not meaningful in this context.

It is, however, necessary to keep in mind that the optimal discrete solution based on the state × time lattice lags behind the optimal solution in the continuous sense. Figure 6.13 has shown how the solution cost varies with the extent of the high-resolution planning region. For scenarios based on real data, the true continuous optimum is very difficult to obtain because this would amount to solving a nonlinear programming (NLP) problem with very complex constraints induced by the unstructured environment

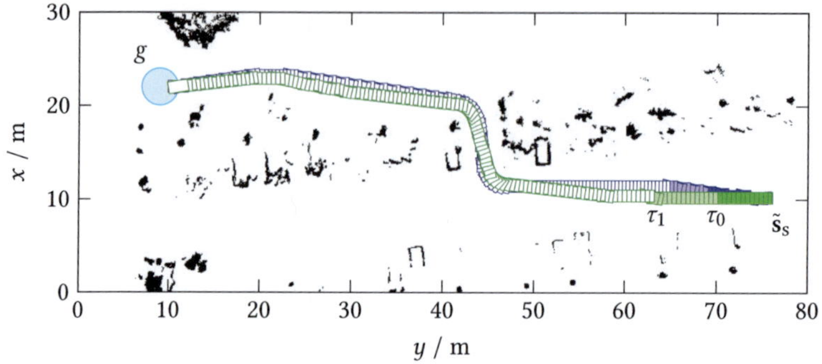

Figure 6.24: Result of pseudo-continuous planning (■□□, $\epsilon = 1$). The solution of the more coarse planning from Figure 6.2 is also shown for comparison (■□□, $\epsilon = 1$). The shadings show dimensionality as usual.

and dynamic obstacles. Yet, in order to provide an assessment for the influence of discretization, the scenario from Figure 6.2 has been replanned with a finer resolution of $\delta^0_{xy} = \delta^1_{xy} = 10\,\text{cm}$ and 96 distinct orientations and with disabled motion primitive decomposition. This is already at the scale of the map representation, which is based on $10\,\text{cm} \times 10\,\text{cm}$ cells. Therefore, it can be assumed that the result (see Figure 6.24) is very close to the continuously optimal solution. The costs decreased only marginally from 74.38 (Figure 6.2) to 73.38 (Figure 6.24), which is further indication that the proposed planning algorithm is able to compute high-quality motion plans.

Obstacle clearance

When planning a path for robotic manipulators, maximum clearance from any obstacle is often desired [Coh12]. However, in the context of mobile robots, this metric only plays a minor role: Unless there is an immediate danger for collision, there is generally no need to deviate from an otherwise optimal trajectory just to increase the distance to the next obstacle. In the proposed planning framework, the observance of the minimum clearance from all obstacles is guaranteed by the modeling of the environment (see Chapter 4) and the incorporation of this model into the planning algorithm.

Robustness in narrow spaces

This metric also originates from planning algorithms based on probabilistic sampling, for which narrow passages are notoriously difficult. Since the proposed planning algorithm is a deterministic search-based planning algorithm, it is very well suited for planning in narrow spaces. Of course, the planning resolution needs to be chosen fine enough to allow the safe passage of narrow corridors.

Smoothness

Since the proposed planning scheme relies on a discretized representation of the state space, the computed motion plans cannot be as smooth as the result of continuous methods are. If desired for a particular task, one can always apply additional trajectory smoothing in a subsequent post-processing step. This is, however, not considered in this thesis since practice has shown that already a simple path controller is sufficient to produce very smooth overall motion on the basis of the computed plan (see Figure 6.21).

Success rate

The evaluation of the success rate also originates from the field of probabilistic sampling-based algorithms. Since those algorithms are generally only *probabilistically complete*, they may fail to find an existing solution. The proposed algorithm, however, is a search-based approach and thus is guaranteed to find a solution if one exists within the chosen discretization. This class of algorithms is called *resolution-complete* [LaV06].

As with virtually all planning algorithms, computation time depends on the environment and the distance to the goal. If computation time is limited, an algorithm may fail to find a solution within the allotted time. For search-based planning algorithms, the computation time depends, besides the implementation details, particularly on the employed graph search algorithm. In this thesis, ARA* was used as it is capable of quickly providing an initial, possibly suboptimal, but valid solution, which is iteratively improved if there is still computation time available in the current cycle. Further details and precise figures were given in Section 6.3.

In addition to the success rate of the planning itself, also the success rate of the actual motion execution needs to be considered. With its combined planning for multiple waypoints, the proposed algorithm can prevent the robot from arriving at an intermediate waypoint in an orientation that might be disadvantageous for the further journey to the final goal or would even lead to the robot getting stuck (see Figure 6.19). This is a unique feature of the proposed planning algorithm.

Occurrence of collisions

The proposed planning algorithm always produces collision free plans (as long as such a plan exists and the future motion of dynamic obstacles is predicted correctly). In this respect, the proposed unified planning of local avoidance maneuvers and global paths may be advantageous over hierarchical methods, which first compute a path without considering dynamic obstacles and then try to modify this initial solution in order to avoid any collision with dynamic obstacles (see Figure 6.15).

Precision for reaching the target

The required precision for reaching the target depends on the particular task. The multi-resolution scheme allows for the specification of a very fine resolution in the vicinity of the goal. This makes it possible to robustly plan complex scenarios like parking in narrow spaces (Figure 6.14) with high precision for the robot's end configuration. On the other hand, waypoint regions may also be specified in a very general manner if high precision is not required. This allows the planning algorithm to determine the optimal state of the robot for reaching each waypoint.

Number of parameters to tune

The parameters needed for the construction of the state × time lattices and the sampling of the corresponding motion primitive sets are listed in Table 6.1. They depend on the particular system model of the robot and need to be chosen in such a way that the robot is able to reach a different lattice point within the duration of a motion primitive. Moreover, the discretization of the high-resolution lattice needs to be fine enough to allow the passage of narrow corridors. The remaining resolution-independent parameters (e.g., maximum quantization error during motion primitive sampling and admissible cost increase for motion primitive decomposition) are relative parameters and the values according to Table 6.1 proved to be sensible in practice. Thus, there will be probably no need to explicitly tune those parameters. Furthermore, additional parameters are required to describe the robot's geometry as well as the desired clearance to obstacles; this is, however, true for any motion planning algorithm.

The choice of the employed cost function is technically independent of the proposed planning algorithm. Nonetheless, a cost function that assesses duration, length and collision risk was provided in this thesis as an example. The choice of the particular weights depends on the user's preferences and the concrete scenario and is thus not the focus of this thesis.

Finally, there are some planning-specific parameters. These are the temporal thresholds τ_i, which were investigated in Figure 6.12, and the extent of the high-resolution planning regions, for which an example has been provided in Figure 6.13.

All in all, the number of parameters to tune is well manageable. Therefore, one can expect a straightforward application of the proposed planning framework to a large variety of robotic platforms and a wide range of scenarios.

6.6 Evaluation: Comparison with Other Algorithms

A comprehensive overview of alternative existing methods has already been given in Chapter 2 providing information on the respective fields of application and the particular advantages and disadvantages. In this section, an additional, more detailed comparison of some selected algorithms will be provided in an attempt to provide a fair comparison of the proposed algorithm with a competitive subset of existing methods. Table 6.3 gives a concise summary of the individual features and properties of each algorithm.

The first column summarizes the properties of sampling-based planning algorithms and underlines the advantages of search-based planning algorithms in the context of mobile robot motion planning. These are first and foremost the resolution-completeness and ability to compute resolution-optimal plans. Since sampling-based planning algorithms have been developed mainly for robotic manipulators with many degrees of free-

Table 6.3: Comparison of algorithms.

	Sampling-based planners (see Section 2.5)	Kinodynamic state lattice planner [Piv08; Piv09c; Lik09]	Time-bounded lattices [Kus09]	Spatiotemporal state lattices for on-road driving [Zie09]	Hybrid A* [Dol08; Mon08; Pet12]	Adaptive state × time lattices [Pet13b; Pet13c; Pet14]
Domain of application						
Dynamic environments	◐	○	●	●	○	●
Unstruct. environments	●	●	●	○	●	●
Rough terrain	○	●	◐	○	●	●
Multiple waypoints	○	○	○	○	◐	●
Planner properties						
Variable dimensionality	○	◐	●	○	○	●
Multiple resolutions	○	◐	○	○	◐	●
Prob. obstacle model	○	○	●	○	○	●
Probabilistic motion primitive sampling	●	○	○	○	○	●
Solution properties						
Kinematically feasible path	●	◐	○	●	●	●
Dynamically feasible trajectory	●	◐	◐	●	○	◐
Resolution-complete	○	●	●	●	●	●
Resolution-optimal	○	●	◐	●	●	●

● applicable ◐ partially applicable ○ not applicable

dom, planning with cost maps, i.e., consideration of terrain characteristics, is generally not considered.

The second column of Table 6.3 addresses the kinodynamic state lattice planner of Pivtoraiko, Knepper, and Kelly [Piv09c]. It is similar to the state lattice planner of Likhachev and Ferguson [Lik09], which was used by *Boss*, the robot that won the DARPA Urban Challenge, and established the successful application of state lattice planners. The method was extended to planning with graduated fidelity in [Piv08]. The authors propose the combination of a four-dimensional high-fidelity lattice and a low-fidelity two-dimensional lattice which corresponds to planning in a simple grid. This makes it impossible to guarantee kinematical feasibility of the complete solution. Planning in a dynamic environment is not considered.

In contrast to the kinodynamic state lattice planner, the time-bounded lattice planner of Kushleyev and Likhachev [Kus09] directly addresses planning in a dynamic environment (Table 6.3, third column). However, as is the case with the former algorithm, no successive dimensionality reduction is employed: at a certain point in time, the search transitions to a search in a simple two-dimensional grid, which neglects the vehicle's kinematics. On the other hand, the algorithm is capable of including (circular) dynamic obstacles using a probabilistic model, which has been the basis for one of the methods adopted in this thesis for the representation of dynamic obstacles.

Column four of Table 6.3 outlines the properties of the spatiotemporal lattice planner of Ziegler and Stiller [Zie09] which has been developed specifically for on-road driving. Due to the restriction of the search space to the road, the algorithm can perform an exhaustive search of the entire lattice and is able to compute high-quality time-parametrized trajectories. While this is desired for on-road driving, it prevents the computation of global plans in unstructured environments.

The three last-mentioned algorithms are all based on planning in state lattices. In order to provide an additional comparison with a non-lattice planner, the proposed planning algorithm is also compared to the *Hybrid A** algorithm of Dolgov et al. [Dol08], which was used by team Stanford's *Junior* in the DARPA Urban Challenge [Mon08]. The properties of Hybrid A* are outlined in Table 6.3, column five. In contrast to state lattice planners, Hybrid A* employs direct discretization of the input space (cf. Figure 3.1). An underlying grid is used to reduce the branching factor of the search graph by limiting the amount of nodes in already visited regions of the search space (see Section 2.6.1). In preparatory work for this thesis, Hybrid A* has been extended with the capability for multi-waypoint planning and was implemented and successfully field-tested on a mobile outdoor robot [Pet12]. Hybrid A*, however, is only concerned with planning kinematically feasible paths; it does not consider dynamic obstacles nor is it capable of planning dynamically feasible trajectories.

The above-mentioned algorithms have been developed with specific applications in mind. As one can see from Table 6.3, none of these algorithms is able to cover all requirements addressed in this thesis. Therefore, a novel highly generic algorithm has been developed to provide a unified framework for computing resolution-optimal motion plans in unstructured dynamic environments. The distinctive feature of the proposed algorithm is the combined planning of local obstacle avoidance maneuvers and global paths while guaranteeing dynamical feasibility of the short-term trajectory and kinematical feasibility of the global path. For this purpose, a novel approach based on time slices for modeling dynamic obstacles with arbitrary shape has been developed in this thesis. The flexible planning frame work allows the straightforward integration of additional features like planning along multiple waypoints or consideration of terrain characteristics. This makes the proposed algorithm usable for a broad range of scenarios.

So far, only the differences regarding the capabilities of the individual algorithms have been discussed. An additional important criterion for the practical applicability is the required planning time of the algorithms. There are mainly two factors that make a fair comparison difficult. On the one hand, the software implementation that has been used to obtain the presented results in the respective papers is generally not publicly available. On the other hand, the used test environments and scenarios are only briefly described in the papers and are hard to reproduce. This is why one needs to rely on the timing results that the authors provide in their publications. For the kinodynamic state lattice planner (Pivtoraiko, Knepper, and Kelly) and the spatiotemporal state lattice planner (Ziegler and Stiller), the authors report computation times in the double-digit millisecond range for typical on-road scenarios; for the time-bounded lattice planner (Kushleyev and Likhachev) and the Hybrid A* planner (Dolgov et al.) planning times in the range of several hundred milliseconds have been reported. In order to provide some reliable figures for the performance comparison of the motion planning algorithm that has been developed in this thesis, an implementation of Hybrid A*, which was built from scratch, has been used to obtain reproducible results for identical problem setups, environments, and robot models. Hybrid A* has been chosen for several reasons. It is well-suited for planning in unstructured environments, it is capable of planning kinematically feasible paths, and it is—with the extension from [Pet12]—able to plan paths via multiple waypoints. In addition, the kinodynamic state lattice planner of Pivtoraiko, Knepper, and Kelly can be considered as a special case of the proposed planning algorithm (i.e., without considering multiple waypoints or dynamic obstacles and thus time-parametrized planning), which is why similar performance can be expected if the proposed algorithm is parametrized accordingly. On the other hand, Hybrid A* is based on the different concept of input space discretization, which makes a comparison much more interesting.

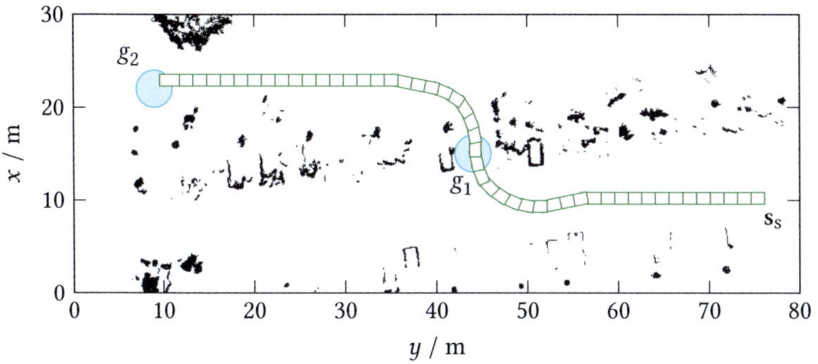

Figure 6.25: Planning with Hybrid A* for the scenario from Figure 6.18. The result is a pure path.

Figure 6.25 shows the result of planning with Hybrid A* for the scenario from Figure 6.18. The computation times stated in [Dol08] could be confirmed: the total planning time was 597 ms. Faster planning times—like the 100 ms mentioned in [Pet12]—can only be achieved for maps with smaller cell count since the size of the underlying grid is coupled with the map resolution. In contrast, planning with the proposed novel algorithm took only 61 ms (see Section 6.3.4). Thus the state × time × goal lattice planner with variable dimensionality and multiple resolutions is a worthy candidate to replace the Hybrid A* planning algorithm. Furthermore, the result of Hybrid A* is a pure path which cannot guarantee dynamic feasibility of the solution nor does it take dynamic obstacles into account. With regard to the optimality of the solution, the resulting path of Hybrid A* is also slightly inferior in the discussed scenario: the path has a total length of 72.0 m while the length of the lattice planner solution from Figure 6.18 is only 71.4 m. On a positive note, the path planned by Hybrid A* is much smoother due to the lack of state space discretization.

Since mobile robot motion planning is a very broad field, the limited scope of a thesis inevitably restricts the comparison to a subset of existing methods. Nonetheless, an attempt was made to select a competitive and representative set of current state-of-the-art algorithms for comparison with the novel approach developed in this thesis. It was shown that the algorithm performs similar to existing state lattice planners with regard to computation time; in addition, it provides further features like simultaneously planning local avoidance maneuvers for dynamic obstacles or motion plans via multiple waypoints. This is made possible by the holistic and generic formulation of the motion planning concept, which provides a sound basis for the integration of additional capabilities.

6.7 Summary

This chapter has presented the results as well as a detailed evaluation and analysis of the proposed planning concept. First, the implementation of the planning algorithm in terms of a portable C++ software library has been described. This implementation has been used to evaluate the planning algorithm both in simulation and on a real robotic platform. Appropriate test scenarios were chosen to highlight specific properties of the planning algorithm. Of course, all features of the planning algorithm were simultaneously active during each planning. The results showed the effective use of successive dimensionality reduction, planning with multiple resolutions, consideration of dynamic obstacles, and combined planning along multiple waypoints. For all presented scenarios[3] the resolution-optimal solution could be computed in—often considerably—less than 100 ms. In field tests, the results of the entire processing pipeline showed that, despite

3. The result of the scenario from Figure 6.2 is an exception since it employed high-resolution planning only and was provided for reference and comparison with the multi-resolution result from Figure 6.8.

the discretizing approach, the planning algorithm enables smooth overall motion of the robot without any intermediate trajectory smoothing. The performance of the proposed planning scheme was discussed on the basis of various evaluation criteria and it has been compared to a competitive set of state-of-the-art motion planning algorithms. A particularly detailed comparison was made with an implementation of the Hybrid A* planning algorithm, which allowed a fair comparison in identical environments and with identical test setups. The proposed planning algorithm was superior to Hybrid A* both in terms of planning time and solution quality. The results presented and discussed in this chapter showed that the hybrid-dimensional multi-resolution state $\times$ time $\times$ goal lattice planner is able to efficiently compute motion plans that combine local obstacle avoidance and global path planning for a wide range of scenarios.

Chapter 7

Conclusions and Future Work

In this thesis, an algorithm for robot motion planning has been developed that allows the joint planning of local obstacle avoidance maneuvers and global, kinematically feasible paths. The theoretic foundations of the planning scheme have been formally stated and the algorithm has been implemented in the form of a portable C++ library. A case study of a mobile robot was used to explain the concept throughout the thesis. After devising a scheme for modeling the environment, a comprehensive evaluation of the planning framework has been conducted, and it was shown that the proposed planning algorithm is capable of computing resolution-optimal motion plans with a rate of approximately 10 Hz that reliably avoid collisions with dynamic obstacles while still optimizing for global optimality of the overall solution.

7.1 Summary of Contributions

The main scientific contribution of this thesis is the rigorous formulation of a holistic planning concept which allows a successive dimensionality reduction and variation of the planning resolution in a consistent way. The planning algorithm produces hybrid solutions which are part trajectory and part path. It is based on state × time lattices, which have been extended to state × time × goal lattices in this thesis to enable the joint planning for multiple waypoints. Furthermore, a generic method has been proposed that probabilistically samples motion primitives, which connect the lattice points, from arbitrary dynamical system models. It was explained how these motion primitives can be used to construct a search graph that converts the motion planning problem to the general problem of finding shortest paths in a graph. For solving the latter, the planning framework proposed in this thesis used the Anytime Repairing A* (ARA*) algorithm, which is capable of quickly finding an initial, possibly suboptimal, but perfectly valid, solution; if there is computation time left for the current planning cycle, ARA* iteratively refines the solution. Since ARA* belongs to the class of heuristic graph search algorithms, appropriate consistent heuristics have been devised, which guide the graph search along the chain of waypoints by estimating the remaining distance to the goal.

Moreover, in order to test the planning algorithm for a wide range of scenarios, it was necessary to develop a method for modeling the environment of the robot—although this was not the focus of this thesis. For this purpose, a scheme has been devised that models the risk for collisions with static and dynamic obstacles in a probabilistic way. An existing method for modeling circular dynamic obstacles has been extended and a novel method based on time slices has been developed for obstacles with arbitrary shape. In addition, consideration of terrain characteristics has been integrated into the environment model. The employed scheme for modeling the environment is only one of many possibilities since the planning algorithm is totally independent of the utilized scheme for collision checking and can thus be easily extended should the need arise.

The proposed planning algorithm was extensively evaluated both in simulation and on a real robotic platform with respect to various criteria. It was shown that the proposed algorithm especially stands out with respect to its capabilities and wide applicability while being at least as fast as other competing algorithms.

7.2 Future Work

The proposed planning concept has been developed with maximum flexibility and extensibility in mind. Therefore, future work might investigate the application of the planning algorithm to other domains. Fore example, it might be interesting to apply the algorithm to multi-waypoint mobile manipulation tasks. These are characterized by many degrees of freedom and therefore constitute a challenging field for search-based planning algorithms. This is where the proposed scheme of successive dimensionality reduction might prove to be beneficial, especially if the motion of the robot is restricted by additional kinematical constraints. In addition, the applicability of the proposed planning algorithm to multi-robot motion planning might be worth investigating (cf. [Fre11; Cir14]).

A further interesting question is whether the proposed planning scheme can profit from advances in multi- and many-core architectures by employing parallelization techniques. The implementation of the motion primitive sampler that has been developed in this thesis already utilizes all available processor cores by sampling all motion primitive bunches in parallel. However, the subsequent search through the lattice is performed sequentially. In the last few years, there have been substantial advances in parallel graph search algorithms (e.g., [Bra12; Phi14; Zho15]), and it might be worthwhile to investigate their applicability and practicability within the proposed planning framework.

Appendix A

Proofs

Proposition 1 (Monotonic cost): *Let* $\tilde{\mathrm{s}}$ *be the currently expanded state during the graph search,* $\tilde{\mathrm{s}}' \in \mathrm{succ}(\tilde{\mathrm{s}})$ *a successor that is reached by the execution of a motion primitive* m*, and* $c(\tilde{\mathrm{s}}, \tilde{\mathrm{s}}')$ *the cost for the transition from* $\tilde{\mathrm{s}}$ *to* $\tilde{\mathrm{s}}'$ *by the application of* m*. If the accumulated costs* $g(\tilde{\mathrm{s}})$ *and* $g(\tilde{\mathrm{s}}')$ *are defined according to* (5.20) *and* (5.21) *with weighting factors* $\eta_t > 0$ *and* $\eta_r \geq 0$*, then* $c(\tilde{\mathrm{s}}, \tilde{\mathrm{s}}') > 0$.

Proof. According to (5.15), the relation $c(\tilde{\mathrm{s}}, \tilde{\mathrm{s}}') = g(\tilde{\mathrm{s}}') - g(\tilde{\mathrm{s}})$ holds. Therefore, in order to prove $c(\tilde{\mathrm{s}}, \tilde{\mathrm{s}}') > 0$, it is sufficient to show that

$$g(\tilde{\mathrm{s}}) < g(\tilde{\mathrm{s}}') \,. \tag{A.1}$$

Substituting (5.20) and (5.21) into (A.1) yields

$$\begin{aligned} &\bar{l}(\tilde{\mathrm{s}}) + \eta_{\mathrm{t}} \bar{t}(\tilde{\mathrm{s}}) + \eta_{\mathrm{r}} \bar{p}_{\mathrm{coll}}(\tilde{\mathrm{s}}) \\ &\quad < \bar{l}(\tilde{\mathrm{s}}) + l(m) + \eta_{\mathrm{t}}\big(\bar{t}(\tilde{\mathrm{s}}) + t(m)\big) \\ &\qquad + \eta_{\mathrm{r}}\Big(1 - \big(1 - \bar{p}_{\mathrm{coll}}(\tilde{\mathrm{s}})\big)\big(1 - p_{\mathrm{coll}}(m)\big)\Big) \,, \end{aligned} \tag{A.2}$$

which can be simplified to the equivalent inequality

$$0 < l(m) + \eta_\mathrm{t}\, t(m) + \eta_\mathrm{r}\, p_\mathrm{coll}(m)\big(1 - \bar{p}_\mathrm{coll}(\tilde{\mathrm{s}})\big)\,. \tag{A.3}$$

The first term of the right-hand side in (A.3) is the length of the motion primitive, $l(m)$. It is by definition a non-negative quantity; however, in case of the "wait" primitive, it may be zero. Thus $l(m) \geq 0$. The second term in (A.3) is the weighted duration of the motion primitive, $\eta_\mathrm{t}\, t(m)$. Since time is always progressing, there is no motion primitive with zero duration, thus $t(m) > 0$ and therefore $\eta_\mathrm{t}\, t(m) > 0$ if and only if $\eta_\mathrm{t} > 0$. The third term in (A.3) is always non-negative for $\eta_\mathrm{r} \geq 0$ because $p_\mathrm{coll}(m)$ and $\bar{p}_\mathrm{coll}(\tilde{\mathrm{s}})$ are probabilities and thus in the closed interval $[0, 1]$. The correctness of (A.3) follows directly from these observations, which concludes the proof. □

Proposition 2 (Consistent heuristics): *Let N be the number of waypoints, $\tilde{\mathrm{s}} = (\tilde{x}, \tilde{y}, \ldots, g_i)$ the currently expanded state during the graph search, $\tilde{\mathrm{s}}' = (\tilde{x}', \tilde{y}', \ldots, g_k) \in \mathrm{succ}(\tilde{\mathrm{s}})$ a successor that is reached by the execution of a motion primitive m, and $c(\tilde{\mathrm{s}}, \tilde{\mathrm{s}}')$ the cost for the transition from $\tilde{\mathrm{s}}$ to $\tilde{\mathrm{s}}'$ by the application of m. The heuristic function $h(\tilde{\mathrm{s}})$ defined according to (5.26) with $\mathfrak{d}(\tilde{\mathrm{s}})$ defined according to (5.27) is consistent.*

Proof. For a heuristic function $h(\tilde{\mathrm{s}})$ to be consistent, the relation

$$h(\tilde{\mathrm{s}}) \leq h(\tilde{\mathrm{s}}') + c(\tilde{\mathrm{s}}, \tilde{\mathrm{s}}') \qquad \forall\, \tilde{\mathrm{s}}, \tilde{\mathrm{s}}' \mid \tilde{\mathrm{s}}' \in \mathrm{succ}(\tilde{\mathrm{s}}) \tag{A.4}$$

must hold [Pea84]. The cost $c(\tilde{\mathrm{s}}, \tilde{\mathrm{s}}')$ can be obtained by substituting (5.20) and (5.21) into (5.15):

$$c(\tilde{\mathrm{s}}, \tilde{\mathrm{s}}') = g(\tilde{\mathrm{s}}') - g(\tilde{\mathrm{s}}) = l(m) + \eta_\mathrm{t}\, t(m) + \underbrace{\eta_\mathrm{r}\, p_\mathrm{coll}(m)\big(1 - \bar{p}_\mathrm{coll}(\tilde{\mathrm{s}})\big)}_{=:p(\tilde{\mathrm{s}}, m)}\,. \tag{A.5}$$

To shorten notation, the expression for the costs induced by the collision risk is briefly denoted by p. Note that $p(\tilde{\mathbf{s}}, m) \geq 0$ since $\eta_\mathrm{r} \geq 0$ and $p_\mathrm{coll}(m), \bar{p}_\mathrm{coll}(\tilde{\mathbf{s}}) \in [0, 1]$. With (5.26) and (5.27), the consistency condition (A.4) becomes

$$\begin{aligned} &\left(1 + \frac{\eta_\mathrm{t}}{\tilde{v}_\mathrm{max}}\right)\left(\mathfrak{d}_i(\tilde{\mathbf{s}}) + \sum_{j=i+1}^{N} \mathfrak{w}_j\right) \\ &\quad \leq \left(1 + \frac{\eta_\mathrm{t}}{\tilde{v}_\mathrm{max}}\right)\left(\mathfrak{d}_k(\tilde{\mathbf{s}}') + \sum_{j=k+1}^{N} \mathfrak{w}_j\right) + l(m) + \eta_\mathrm{t}\, t(m) + p(\tilde{\mathbf{s}}, m)\,. \end{aligned} \tag{A.6}$$

In order to prove the correctness of (A.6), two cases need to be considered: a) $\tilde{\mathbf{s}} = (\tilde{x}, \tilde{y}, \ldots, g_i)$ and $\tilde{\mathbf{s}}' = (\tilde{x}', \tilde{y}', \ldots, g_k)$ belong to the same planning segment, i.e., $g_k = g_i$; b) $\tilde{\mathbf{s}}$ and $\tilde{\mathbf{s}}'$ belong to different planning segments, i.e., $g_k = g_{i+1}$, because waypoint g_i has been reached by $\tilde{\mathbf{s}}'$. For the sake of clarity, the abbreviations

$$\mathbf{s} := \begin{bmatrix} \tilde{x} \\ \tilde{y} \end{bmatrix}, \qquad \mathbf{s}' := \begin{bmatrix} \tilde{x}' \\ \tilde{y}' \end{bmatrix}, \qquad \mathbf{s}_{g_i} := \begin{bmatrix} x_{g_i} \\ y_{g_i} \end{bmatrix} \tag{A.7}$$

are used in the remainder of this proof since no confusion may arise in this context.

a) same planning segment: $g_k = g_i$. With $k = i$, the two sums in (A.6) are equivalent and can thus be subtracted from both sides of the inequality:

$$\left(1 + \frac{\eta_\mathrm{t}}{\tilde{v}_\mathrm{max}}\right)\mathfrak{d}_i(\tilde{\mathbf{s}}) \leq \left(1 + \frac{\eta_\mathrm{t}}{\tilde{v}_\mathrm{max}}\right)\mathfrak{d}_i(\tilde{\mathbf{s}}') + l(m) + \eta_\mathrm{t}\, t(m) + p(\tilde{\mathbf{s}}, m)\,. \tag{A.8}$$

With definition (5.28) for the heuristic distance $\mathfrak{d}_i(\tilde{\mathbf{s}})$ and the abbreviations from (A.7), (A.8) can be written as

$$\left(1+\frac{\eta_\mathrm{t}}{\tilde{v}_\mathrm{max}}\right)\max\left(0, \|\mathbf{s}-\mathbf{s}_{g_i}\|-R_{g_i}\right)$$
$$\leq \left(1+\frac{\eta_\mathrm{t}}{\tilde{v}_\mathrm{max}}\right)\max\left(0, \|\mathbf{s}'-\mathbf{s}_{g_i}\|-R_{g_i}\right) + l(m) + \eta_\mathrm{t}\, t(m) + p(\tilde{\mathbf{s}}, m)\,, \tag{A.9}$$

where $\|\cdot\|$ is the Euclidean norm of a vector. As the left-hand side attains its maximum value for $\|\mathbf{s}-\mathbf{s}_{g_i}\|-R_{g_i} \geq 0$ and the right-hand side attains its minimum value for $\|\mathbf{s}'-\mathbf{s}_{g_i}\|-R_{g_i} \leq 0$, it is sufficient to consider this particular case and prove

$$\left(1+\frac{\eta_\mathrm{t}}{\tilde{v}_\mathrm{max}}\right)\left(\|\mathbf{s}-\mathbf{s}_{g_i}\|-R_{g_i}\right) \leq l(m) + \eta_\mathrm{t}\, t(m) + p(\tilde{\mathbf{s}}, m)\,. \tag{A.10}$$

The proof starts with the left-hand side and makes use of the average velocity $v(m) = l(m)/t(m)$, $0 \leq v(m) \leq \tilde{v}_\mathrm{max}$ of the motion primitive m.

$$\left(1+\frac{\eta_\mathrm{t}}{\tilde{v}_\mathrm{max}}\right)\left(\|\mathbf{s}-\mathbf{s}_{g_i}\|-R_{g_i}\right) = \left(1+\frac{\eta_\mathrm{t}}{\tilde{v}_\mathrm{max}}\right)\left(\|\mathbf{s}-\mathbf{s}_{g_i}+\mathbf{s}'-\mathbf{s}'\|-R_{g_i}\right)$$
$$\leq \left(1+\frac{\eta_\mathrm{t}}{\tilde{v}_\mathrm{max}}\right)\left(\|\mathbf{s}-\mathbf{s}'\| + \underbrace{\|\mathbf{s}'-\mathbf{s}_{g_i}\|-R_{g_i}}_{\leq 0 \text{ by assumption}}\right) \leq \left(1+\frac{\eta_\mathrm{t}}{\tilde{v}_\mathrm{max}}\right)\|\mathbf{s}-\mathbf{s}'\|$$
$$\leq \left(1+\frac{\eta_\mathrm{t}}{\tilde{v}_\mathrm{max}}\right) l(m) \leq \left(1+\frac{\eta_\mathrm{t}}{v(m)}\right) l(m) = l(m) + \eta_\mathrm{t}\, t(m)$$
$$\leq l(m) + \eta_\mathrm{t}\, t(m) + p(\tilde{\mathbf{s}}, m)\,, \tag{A.11}$$

which is the right-hand side of (A.10) and thus concludes the proof for the case $g_k = g_i$.

b) different planning segments: $g_k = g_{i+1}$. With $k = i + 1$, inequality (A.6) becomes

$$\left(1+\frac{\eta_\mathrm{t}}{\tilde{v}_{\max}}\right)\left(\mathfrak{d}_i(\tilde{\mathbf{s}}) + \sum_{j=i+1}^{N} \mathfrak{w}_j\right) \leq \left(1+\frac{\eta_\mathrm{t}}{\tilde{v}_{\max}}\right)\left(\mathfrak{d}_{i+1}(\tilde{\mathbf{s}}') + \sum_{j=i+2}^{N} \mathfrak{w}_j\right) + l(m) + \eta_\mathrm{t}\, t(m) + p(\tilde{\mathbf{s}}, m)\,, \tag{A.12}$$

which can be simplified to

$$\left(1+\frac{\eta_\mathrm{t}}{\tilde{v}_{\max}}\right)\left(\mathfrak{d}_i(\tilde{\mathbf{s}}) + \mathfrak{w}_{i+1}\right) \leq \left(1+\frac{\eta_\mathrm{t}}{\tilde{v}_{\max}}\right)\mathfrak{d}_{i+1}(\tilde{\mathbf{s}}') + l(m) + \eta_\mathrm{t}\, t(m) + p(\tilde{\mathbf{s}}, m)\,. \tag{A.13}$$

With definition (5.28) for the heuristic distance $\mathfrak{d}_i(\tilde{\mathbf{s}})$, definition (5.29) for the inter-waypoint distance $\mathfrak{w}_j$, and the abbreviations from (A.7), (A.13) can be written as

$$\left(1+\frac{\eta_\mathrm{t}}{\tilde{v}_{\max}}\right)\left(\max\left(0, \|\mathbf{s} - \mathbf{s}_{g_i}\| - R_{g_i}\right) + \max\left(0, \|\mathbf{s}_{g_{i+1}} - \mathbf{s}_{g_i}\| - R_{g_{i+1}} - R_{g_i}\right)\right) \leq \left(1+\frac{\eta_\mathrm{t}}{\tilde{v}_{\max}}\right) \max\left(0, \|\mathbf{s}' - \mathbf{s}_{g_{i+1}}\| - R_{g_{i+1}}\right) + l(m) + \eta_\mathrm{t}\, t(m) + p(\tilde{\mathbf{s}}, m)\,. \tag{A.14}$$

The left-hand side attains its maximum value for $\|\mathbf{s} - \mathbf{s}_{g_i}\| - R_{g_i} \geq 0$ and $\|\mathbf{s}_{g_{i+1}} - \mathbf{s}_{g_i}\| - R_{g_{i+1}} - R_{g_i} \geq 0$; the right-hand side attains its minimum value for $\|\mathbf{s}' - \mathbf{s}_{g_{i+1}}\| - R_{g_{i+1}} \leq 0$. Thus it is sufficient to consider this particular case and prove

$$\left(1+\frac{\eta_\mathrm{t}}{\tilde{v}_{\max}}\right)\left(\|\mathbf{s} - \mathbf{s}_{g_i}\| - R_{g_i} + \|\mathbf{s}_{g_{i+1}} - \mathbf{s}_{g_i}\| - R_{g_{i+1}} - R_{g_i}\right) \leq l(m) + \eta_\mathrm{t}\, t(m) + p(\tilde{\mathbf{s}}, m)\,. \tag{A.15}$$

Again, the proof starts with the left-hand side and makes use of the average velocity $v(m) = l(m)/t(m)$, $0 \leq v(m) \leq \tilde{v}_{\max}$ of the motion primitive m. Additionally, the relation

$$\|\mathbf{s}' - \mathbf{s}_{g_i}\| - R_{g_i} \leq 0 \tag{A.16}$$

holds because $\mathbf{s}'$ caused a transition to the next planning segment and thus must lie inside waypoint g_i, i.e., $\tilde{\mathbf{s}}' \in \tilde{\mathcal{S}}_{g_i}$ according to (5.11) and (5.23).

$$\begin{aligned}
&\left(1 + \frac{\eta_\mathrm{t}}{\tilde{v}_{\max}}\right)\left(\|\mathbf{s} - \mathbf{s}_{g_i}\| - R_{g_i} + \|\mathbf{s}_{g_{i+1}} - \mathbf{s}_{g_i}\| - R_{g_{i+1}} - R_{g_i}\right) \\
&\quad = \left(1 + \frac{\eta_\mathrm{t}}{\tilde{v}_{\max}}\right)\Big(\|\mathbf{s} - \mathbf{s}_{g_i} + \mathbf{s}' - \mathbf{s}'\| - R_{g_i} \\
&\qquad\qquad + \|\mathbf{s}_{g_{i+1}} - \mathbf{s}_{g_i} + \mathbf{s}' - \mathbf{s}'\| - R_{g_{i+1}} - R_{g_i}\Big) \\
&\quad \leq \left(1 + \frac{\eta_\mathrm{t}}{\tilde{v}_{\max}}\right)\Big(\|\mathbf{s} - \mathbf{s}'\| + \underbrace{\|\mathbf{s}' - \mathbf{s}_{g_i}\| - R_{g_i}}_{\leq 0 \text{ by (A.16)}} \\
&\qquad\qquad + \underbrace{\|\mathbf{s}' - \mathbf{s}_{g_i}\| - R_{g_i}}_{\leq 0 \text{ by (A.16)}} + \underbrace{\|\mathbf{s}' - \mathbf{s}_{g_{i+1}}\| - R_{g_{i+1}}}_{\leq 0 \text{ by assumption}}\Big) \\
&\quad \leq \left(1 + \frac{\eta_\mathrm{t}}{\tilde{v}_{\max}}\right)\|\mathbf{s} - \mathbf{s}'\| \leq \left(1 + \frac{\eta_\mathrm{t}}{\tilde{v}_{\max}}\right) l(m) \\
&\quad \leq \left(1 + \frac{\eta_\mathrm{t}}{v(m)}\right) l(m) = l(m) + \eta_\mathrm{t}\, t(m) \\
&\quad \leq l(m) + \eta_\mathrm{t}\, t(m) + p(\tilde{\mathbf{s}}, m)\,,
\end{aligned} \tag{A.17}$$

which is the right-hand side of (A.15) and thus concludes the proof for the case $g_k = g_{i+1}$. □

Appendix B

Algorithms

This appendix provides the pseudocode for two major components of the proposed motion planning algorithm, namely the sampling of motion primitive bunches and the Anytime Repairing A* (ARA*) graph search algorithm. The entire motion planning algorithm has been implemented in C++ to guarantee fast execution and easy portability (see Section 6.1).

Sampling of Motion Primitive Bunches

Sampling motion primitive sets from arbitrary system models has been introduced in Section 3.5.4. Algorithm 1 describes the procedure for sampling a single motion primitive bunch $\mathcal{B}(\tilde{\mathbf{s}}_0)$ starting at the respective lattice point $\tilde{\mathbf{s}}_0$. Since the algorithm has no side effects, the sampling of all motion primitive bunches can be performed in parallel.

Algorithm 1 Sampling of a motion primitive bunch.

Inputs:

$\tilde{\mathbf{s}}_0$	Start state of bunch
N_{total}	Total number of samples
N_{expl}	Number of samples during exploration phase
$e_{\text{q,max}}$	Maximum quantization error
$n_{\max}$	Maximum number of time intervals for a motion primitive
δ_t	Time increment

procedure SampleBunch($\tilde{\mathbf{s}}_0, N_{\text{total}}, N_{\text{expl}}, e_{\text{q,max}}, n_{\max}, \delta_t$)
  $\mathcal{B}(\tilde{\mathbf{s}}_0) \leftarrow \varnothing$
  for $i \leftarrow 1$ **to** N_{total} **do**
    $\mathcal{T} \leftarrow ()$
    for $k \leftarrow 1$ **to** $n_{\max}$ **do**
      $\mathbf{u}_k(t) \leftarrow$ SampleInput(t_{k-1}, t_k) ▹ see (3.47)
      $\mathbf{u}(t) \leftarrow \sum_{j=1}^{k} \mathbf{u}_j(t)$
      $\mathbf{s}_\text{e} \leftarrow \Phi(\tilde{\mathbf{s}}_0, 0, \mathbf{u}, k\delta_t)$ ▹ see (3.36)
      $\mathcal{T} \leftarrow \left(\mathcal{T}, (\mathbf{s}_\text{e}, k\delta_t)\right)$
      break if trajectory $\mathcal{T}$ violates state constraints
      if $e_\text{q}(\mathbf{s}_\text{e}) < e_{\text{q,max}}$ **then** ▹ see (3.8)
        $\tilde{\mathbf{s}}_\text{e} \leftarrow \boldsymbol{\lambda}(\mathbf{s}_\text{e})$ ▹ see (3.6)
        $m \leftarrow (\tilde{\mathbf{s}}_0, \tilde{\mathbf{s}}_\text{e}, \mathbf{u}, \mathcal{T}, k\delta_t)$
        if $\exists m' = (\tilde{\mathbf{s}}_0, \tilde{\mathbf{s}}'_\text{e}, \mathbf{u}', \mathcal{T}', \Delta t'_\text{m}) \in \mathcal{B}(\tilde{\mathbf{s}}_0) : (\tilde{\mathbf{s}}_\text{e} = \tilde{\mathbf{s}}'_\text{e})$ **then**
          if $J_\text{q}(m) < J_\text{q}(m')$ **then** ▹ see (3.50)
            $\mathcal{B}(\tilde{\mathbf{s}}_0) \leftarrow \mathcal{B}(\tilde{\mathbf{s}}_0) \setminus \{ m' \} \cup \{ m \}$
          end if
        else if $i \leq N_{\text{expl}}$ **then**
          $\mathcal{B}(\tilde{\mathbf{s}}_0) \leftarrow \mathcal{B}(\tilde{\mathbf{s}}_0) \cup \{ m \}$
        end if
      end if
    end for
  end for
  return $\mathcal{B}(\tilde{\mathbf{s}}_0)$
end procedure

Anytime Repairing A* (ARA*)

The Anytime Repairing A* (ARA*) algorithm was proposed in [Lik03a] and is cited here for reference. Due to its anytime nature, ARA* is very well suited for planning fast avoidance maneuvers in a dynamic environment (see Section 5.2.2), which is why it is used extensively in this thesis for solving the eventual graph search problem.

Algorithm 2 Anytime Repairing A* (ARA*) [Lik03a]

```
procedure FVALUE(s̃)
    return g(s̃) + ε h(s̃)                                  ▹ see (5.13)
end procedure

procedure IMPROVEPATH
    while FVALUE(s̃_goal) > min_{s̃∈OPEN} FVALUE(s̃) do
        remove s̃ with the smallest FVALUE(s̃) from OPEN
        CLOSED ← CLOSED ∪ { s̃ }
        for each successor s̃′ of s̃ do                    ▹ see (5.10)
            if s̃′ was not visited before then
                g(s̃′) ← ∞
            end if
            if g(s̃′) > g(s̃) + c(s̃, s̃′) then
                g(s̃′) ← g(s̃) + c(s̃, s̃′)                  ▹ see (5.15)
                if s̃′ ∉ CLOSED then
                    insert s̃′ into OPEN with FVALUE(s̃′)
                else
                    insert/update s̃′ in INCONS            ▹ see [Lik08]
                end if
            end if
        end for
    end while
end procedure
```

Algorithm 2 (Continued) Anytime Repairing A* (ARA*) [Lik03a]

procedure MAIN
    $g(\tilde{\mathbf{s}}_{\text{goal}}) \leftarrow \infty$
    $g(\tilde{\mathbf{s}}_{\text{start}}) \leftarrow 0$
    $\epsilon \leftarrow \epsilon_{\text{start}}$
    $\mathit{OPEN} \leftarrow \mathit{CLOSED} \leftarrow \mathit{INCONS} \leftarrow \emptyset$
    insert $\tilde{\mathbf{s}}_{\text{start}}$ into *OPEN* with FVALUE($\tilde{\mathbf{s}}_{\text{start}}$)
    IMPROVEPATH
    $\epsilon' \leftarrow \min\left(\epsilon,\ g(\tilde{\mathbf{s}}_{\text{goal}}) / \min_{\tilde{\mathbf{s}} \in \mathit{OPEN} \cup \mathit{INCONS}}(g(\tilde{\mathbf{s}}) + h(\tilde{\mathbf{s}}))\right)$ ▷ see (5.18)
    publish current ϵ'-suboptimal solution
    while $\epsilon' > 1$ **do**
        decrease ϵ
        move states from *INCONS* into *OPEN*
        update the priorities for all $\tilde{\mathbf{s}} \in \mathit{OPEN}$ according to FVALUE($\tilde{\mathbf{s}}$)
        $\mathit{CLOSED} \leftarrow \emptyset$
        IMPROVEPATH
        $\epsilon' \leftarrow \min\left(\epsilon,\ g(\tilde{\mathbf{s}}_{\text{goal}}) / \min_{\tilde{\mathbf{s}} \in \mathit{OPEN} \cup \mathit{INCONS}}(g(\tilde{\mathbf{s}}) + h(\tilde{\mathbf{s}}))\right)$
        publish current ϵ'-suboptimal solution
    end while
end procedure

Bibliography

[Agh13] F. Aghili and A. Salerno. "Driftless 3-D attitude determination and positioning of mobile robots by integration of IMU with two RTK GPSs." In: *IEEE/ASME Transactions on Mechatronics* 18.1 (2013), pp. 21–31.

[Bak04] C. Baker, A. Morris, D. Ferguson, S. Thayer, C. Whittaker, Z. Omohundro, C. Reverte, W. Whittaker, D. Hahnel, and S. Thrun. "A campaign in autonomous mine mapping." In: *Proceedings of the IEEE International Conference on Robotics and Automation.* 2004, pp. 2004–2009.

[Bal00] J. Baltes. "A benchmark suite for mobile robots." In: *Proceedings of the IEEE/RSJ International Conference on Intelligent Robots and Systems.* 2000.

[Bar89] J. Barraquand and J.-C. Latombe. "On nonholonomic mobile robots and optimal maneuvering." In: *Proceedings of the IEEE International Symposium on Intelligent Control.* 1989, pp. 340–347.

[Bar91a] J. Barraquand and J.-C. Latombe. "Nonholonomic multibody mobile robots: Controllability and motion planning in the presence of obstacles." In: *Proceedings of the IEEE International Conference on Robotics and Automation.* 1991, pp. 2328–2335.

[Bar91b] J. Barraquand and J.-C. Latombe. "Robot motion planning: a distributed representation approach." In: *The International Journal of Robotics Research* 10.6 (1991), pp. 628–649.

[Bar93] J. Barraquand and J.-C. Latombe. "Nonholonomic multibody mobile robots: Controllability and motion planning in the presence of obstacles." In: *Algorithmica* 10.2–4 (1993), pp. 121–155.

[Ber06] J. van den Berg, D. Ferguson, and J. Kuffner. "Anytime path planning and replanning in dynamic environments." In: *Proceedings of the IEEE International Conference on Robotics and Automation.* 2006, pp. 2366–2371.

[Bet98] J. T. Betts. "Survey of numerical methods for trajectory optimization." In: *AIAA Journal of Guidance, Control, and Dynamics* 21.2 (1998), pp. 193–207.

[Bic02] A. Bicchi, A. Marigo, and B. Piccoli. "On the Reachability of Quantized Control Systems." In: *IEEE Transactions on Automatic Control* 47.4 (2002), pp. 546–563.

[Ble11a] T. Blechmann. *Boost.Heap.* 2011. URL: `http://www.boost.org/doc/libs/1_60_0/doc/html/heap.html` (retrieved on Feb. 4, 2016).

[Ble11b] T. Blechmann. *Boost.Heap. Data Structures.* 2011. URL: `http://www.boost.org/doc/libs/1_60_0/doc/html/heap/data_structures.html` (retrieved on Feb. 4, 2016).

[Boh00] R. Bohlin and L. E. Kavraki. "Path planning using Lazy PRM." In: *Proceedings of the IEEE International Conference on Robotics and Automation.* 2000, pp. 521–528.

[Boh01] R. Bohlin. "Path planning in practice; lazy evaluation on a multi-resolution grid." In: *Proceedings of the IEEE/RSJ International Conference on Intelligent Robots and Systems.* 2001, pp. 49–54.

[Bra01] M. S. Branicky, S. M. LaValle, K. Olson, and L. Yang. "Quasi-randomized path planning." In: *Proceedings of the IEEE International Conference on Robotics and Automation.* 2001, pp. 1481–1487.

[Bra08] M. S. Branicky, R. A. Knepper, and J. J. Kuffner. "Path and trajectory diversity: Theory and algorithms." In: *Proceedings of the IEEE International Conference on Robotics and Automation.* 2008, pp. 1359–1364.

[Bra12] S. Brand and R. Bidarra. "Multi-core scalable and efficient pathfinding with Parallel Ripple Search." In: *Computer Animation and Virtual Worlds* 23.2 (2012), pp. 73–85.

[Bro99] O. Brock and O. Khatib. "High-speed navigation using the global dynamic window approach." In: *Proceedings of the IEEE International Conference on Robotics and Automation.* 1999, pp. 341–346.

[Bue08] M. Buehler, K. Iagnemma, and S. Singh, eds. *Journal of Field Robotics* 25.8–10 (2008): *Special Issue on the 2007 DARPA Urban Challenge, Part I–III.*

[Buh08] J. P. Buhler and P. Stevenhagen. *Algorithmic Number Theory. Lattices, Number Fields, Curves and Cryptography.* Mathematical Sciences Research Institute Publications 44. Cambridge: Cambridge University Press, 2008.

[Cal08] D. Calisi, L. Iocchi, and D. Nardi. "A unified benchmark framework for autonomous mobile robots and vehicles motion algorithms (MoVeMA benchmarks)." In: *Workshop on experimental methodology and benchmarking in robotics research.* 2008.

[Cha87] B. Chazelle. "Approximation and decomposition of shapes." In: *Algorithmic and Geometric Aspects of Robotics.* Ed. by J. T. Schwartz and C.-K. Yap. Hillsdale, NJ: L. Erlbaum Associates, 1987, pp. 145–185.

[Che02] P. Cheng and S. M. LaValle. "Resolution complete rapidly-exploring random trees." In: *Proceedings of the IEEE International Conference on Robotics and Automation.* 2002, pp. 267–272.

[Che99] M. Cherif. "Kinodynamic motion planning for all-terrain wheeled vehicles." In: *Proceedings of the IEEE International Conference on Robotics and Automation.* 1999, pp. 317–322.

[Chi12] S. Chitta, I. A. Şucan, and S. Cousins. "MoveIt!" In: *IEEE Robotics & Automation Magazine* 19.1 (2012). URL: `http://moveit.ros.org`.

[Cho05] H. Choset, K. M. Lynch, S. Hutchinson, G. Kantor, W. Burgard, L. E. Kavraki, and S. Thrun. *Principles of Robot Motion: Theory, Algorithms, and Implementation.* Cambridge, MA: MIT Press, 2005.

[Cir14] M. Cirillo, T. Uras, and S. Koenig. "A lattice-based approach to multi-robot motion planning for non-holonomic vehicles." In: *Proceedings of the IEEE/RSJ International Conference on Intelligent Robots and Systems.* 2014, pp. 232–239.

[Coh10] B. J. Cohen, S. Chitta, and M. Likhachev. "Search-based planning for manipulation with motion primitives." In: *Proceedings of the IEEE International Conference on Robotics and Automation.* 2010, pp. 2902–2908.

[Coh11] B. J. Cohen, G. Subramanian, S. Chitta, and M. Likhachev. "Planning for manipulation with adaptive motion primitives." In: *Proceedings of the IEEE International Conference on Robotics and Automation.* 2011, pp. 5478–5485.

[Coh12] B. Cohen, I. A. Şucan, and S. Chitta. "A generic infrastructure for benchmarking motion planners." In: *Proceedings of the IEEE/RSJ International Conference on Intelligent Robots and Systems.* 2012, pp. 589–595.

[Cor09] T. H. Cormen, C. E. Leiserson, R. L. Rivest, and C. Stein. *Introduction to Algorithms.* 3rd ed. Cambridge, MA: MIT Press, 2009.

[Cur15] F. Curatella, P. Vinetti, G. Rizzo, T. Vladimirova, L. D. Vendictis, T. Emter, J. Petereit, C. W. Frey, D. Usher, I. Stanciugelu, J. Schaefer, E. den Breejen, L. Gisslén, and D. Letalick. "Toward a multifaceted platform for humanitarian demining." In: *13th IARP Workshop on Humanitarian Demining and Risky Intervention.* 12th International Symposium "MINE ACTION 2015". 2015, pp. 133–144.

[DeL98] A. De Luca, G. Oriolo, and C. Samson. "Feedback control of a nonholonomic car-like robot." In: *Robot Motion Planning and Control.* Ed. by J.-P. Laumond. Lecture Notes in Control and Information Sciences 229. Berlin: Springer, 1998, pp. 171–253.

[Dij59] E. W. Dijkstra. "A note on two problems in connexion with graphs." In: *Numerische Mathematik* 1.1 (1959), pp. 269–271.

[Dol08] D. Dolgov, S. Thrun, M. Montemerlo, and J. Diebel. "Practical search techniques in path planning for autonomous driving." In: *Proceedings of the First International Symposium on Search Techniques in Artificial Intelligence and Robotics*. 2008.

[Dol10] D. Dolgov, S. Thrun, M. Montemerlo, and J. Diebel. "Path planning for autonomous vehicles in unknown semi-structured environments." In: *The International Journal of Robotics Research* 29.5 (2010), pp. 485–501.

[Don93] B. Donald, P. Xavier, J. Canny, and J. Reif. "Kinodynamic motion planning." In: *Journal of the Association for Computing Machinery* 40.5 (1993), pp. 1048–1066.

[Dub57] L. E. Dubins. "On curves of minimal length with a constraint on average curvature, and with prescribed initial and terminal positions and tangents." In: *American Journal of Mathematics* 79.3 (1957), pp. 497–516.

[DuT10] N. E. Du Toit and J. W. Burdick. "Robotic motion planning in dynamic, cluttered, uncertain environments." In: *Proceedings of the IEEE International Conference on Robotics and Automation*. 2010, pp. 966–973.

[Emt10] T. Emter, A. Saltoğlu, and J. Petereit. "Multi-sensor fusion for localization of a mobile robot in outdoor environments." In: *Proceedings of the 41st International Symposium on Robotics / 6th German Conference on Robotics (ISR/ROBOTIK 2010)*. 2010.

[Emt12] T. Emter and T. Ulrich. "Fusion of geometrical and visual information for localization and mapping in outdoor environments." In: *Ubiquitous Positioning, Indoor Navigation, and Location Based Service (UPINLBS)*. 2012, pp. 1–5.

[Emt14] T. Emter and J. Petereit. "Integrated multi-sensor fusion for mapping and localization in outdoor environments for mobile robots." In: *Proceedings of SPIE, volume 9121. Multisensor, Multisource Information Fusion: Architectures, Algorithms, and Applications*. 2014.

[Eri09] L. H. Erickson and S. M. LaValle. "Survivability: Measuring and ensuring path diversity." In: *Proceedings of the IEEE International Conference on Robotics and Automation.* 2009, pp. 2068–2073.

[Fer05] D. Ferguson, M. Likhachev, and A. Stentz. "A guide to heuristic-based path planning." In: *Proceedings of the International Conference on Automated Planning and Scheduling.* 2005.

[Fer06a] D. Ferguson and A. Stentz. "Anytime RRTs." In: *Proceedings of the IEEE/RSJ International Conference on Intelligent Robots and Systems.* 2006, pp. 5369–5375.

[Fer06b] D. Ferguson and A. Stentz. "Using interpolation to improve path planning: the field D* algorithm." In: *Journal of Field Robotics* 23.2 (2006), pp. 79–101.

[Fer08a] D. Ferguson, T. M. Howard, and M. Likhachev. "Motion planning in urban environments." In: *Journal of Field Robotics* 25.11–12 (2008), pp. 939–960.

[Fer08b] D. Ferguson and M. Likhachev. "Efficiently using cost maps for planning complex maneuvers." In: *IEEE International Conference on Robotics and Automation: Workshop on Planning with Cost Maps.* 2008.

[Fer98] P. Ferbach. "A method of progressive constraints for nonholonomic motion planning." In: *IEEE Transactions on Robotics and Automation* 14.1 (1998), pp. 172–179.

[Fio98] P. Fiorini and Z. Shiller. "Motion planning in dynamic environments using velocity obstacles." In: *The International Journal of Robotics Research* 17.7 (1998), pp. 760–772.

[Fox97] D. Fox, W. Burgard, and S. Thrun. "The dynamic window approach to collision avoidance." In: *IEEE Robotics & Automation Magazine* 4.1 (1997), pp. 23–33.

[Fra02] E. Frazzoli, M. A. Dahleh, and E. Feron. "Real-time motion planning for agile autonomous vehicles." In: *AIAA Journal of Guidance, Control, and Dynamics* 25.1 (2002), pp. 116–129.

[Fra04a] T. Fraichard and A. Scheuer. "From Reeds and Shepp's to continuous-curvature paths." In: *IEEE Transactions on Robotics* 20.6 (2004), pp. 1025–1035.

[Fra04b] T. Fraichard and H. Asama. "Inevitable collision states – a step towards safer robots." In: *Advanced Robotics* 18.10 (2004), pp. 1001–1024.

[Fra05] E. Frazzoli, M. A. Dahleh, and E. Feron. "Maneuver-based motion planning for nonlinear systems with symmetries." In: *IEEE Transactions on Robotics* 21.6 (2005), pp. 1077–1091.

[Fra93] T. Fraichard. "Dynamic trajectory planning with dynamic constraints: A 'state-time space' approach." In: *Proceedings of the IEEE/RSJ International Conference on Intelligent Robots and Systems.* 1993, pp. 1393–1400.

[Fre11] C. Frese and J. Beyerer. "A comparison of motion planning algorithms for cooperative collision avoidance of multiple cognitive automobiles." In: *Proceedings of the IEEE Intelligent Vehicles Symposium.* 2011, pp. 1156–1162.

[Fre15] C. Frey, T. Emter, J. Petereit, I. Tchouchenkov, T. Müller, M. Tittel, R. Worst, K. Pfeiffer, M. Walter, J. Wöllenstein, S. Rademacher, A. Wenzel, and F. Müller. "Situation responsive networking of mobile robots for disaster management." In: *Proceedings of the IEEE International Symposium on Technologies for Homeland Security.* 2015.

[Ful08] C. Fulgenzi, C. Tay, A. Spalanzani, and C. Laugier. "Probabilistic navigation in dynamic environment using Rapidly-exploring Random Trees and Gaussian processes." In: *Proceedings of the IEEE/RSJ International Conference on Intelligent Robots and Systems.* 2008, pp. 1056–1062.

[Ger08] B. P. Gerkey and K. Konolige. "Planning and control in unstructured terrain." In: *IEEE International Conference on Robotics and Automation: Workshop on Planning with Cost Maps.* 2008.

[Goc11] K. Gochev, B. J. Cohen, J. Butzke, A. Safonova, and M. Likhachev. "Path planning with adaptive dimensionality." In: *Proceedings of the Fourth International Symposium on Combinatorial Search*. 2011.

[Goc12] K. Gochev, A. Safonova, and M. Likhachev. "Planning with adaptive dimensionality for mobile manipulation." In: *Proceedings of the IEEE International Conference on Robotics and Automation*. 2012, pp. 2944–2951.

[Goc13] K. Gochev, A. Safonova, and M. Likhachev. "Incremental planning with adaptive dimensionality." In: *Proceedings of the Twenty-Third International Conference on Automated Planning and Scheduling*. 2013.

[Gon12] J. P. Gonzalez, A. Dornbush, and M. Likhachev. "Using state dominance for path planning in dynamic environments with moving obstacles." In: *Proceedings of the IEEE International Conference on Robotics and Automation*. 2012, pp. 4009–4015.

[Gre07] C. J. Green and A. Kelly. "Toward optimal sampling in the space of paths." In: *13th International Symposium of Robotics Research*. 2007.

[Gri05] G. Grisetti, C. Stachniss, and W. Burgard. "Improving grid-based SLAM with Rao-Blackwellized particle filters by adaptive proposals and selective resampling." In: *Proceedings of the IEEE International Conference on Robotics and Automation*. 2005, pp. 2432–2437.

[Gua15] T. Guan and C. W. Frey. "Reuse historic costs in dynamic programming to reduce computational complexity in the context of model predictive optimization." In: *IEEE International Conference on Vehicular Electronics and Safety*. 2015, pp. 256–263.

[Har68] P. E. Hart, N. J. Nilsson, and B. Raphael. "A formal basis for the heuristic determination of minimum cost paths." In: *IEEE Transactions on Systems Science and Cybernetics* 4.2 (1968), pp. 100–107.

[Har87] R. M. Haralick, S. R. Sternberg, and X. Zhuang. "Image analysis using mathematical morphology." In: *IEEE Transactions on Pattern Analysis and Machine Intelligence* 9.4 (1987), pp. 532–550.

[Her14] C. Hernández, J. A. Baier, and R. Asín. “Making A* run faster than D*-Lite for path-planning in partially known terrain.” In: *Proceedings of the 24th International Conference on Automated Planning and Scheduling*. 2014.

[How07] T. M. Howard and A. Kelly. “Optimal rough terrain trajectory generation for wheeled mobile robots.” In: *The International Journal of Robotics Research* 26.5 (2007), pp. 141–166.

[How08] T. M. Howard, C. J. Green, A. Kelly, and D. Ferguson. “State space sampling of feasible motions for high-performance mobile robot navigation in complex environments.” In: *Journal of Field Robotics* 25.6–7 (2008), pp. 325–345.

[How10] T. M. Howard, C. J. Green, and A. Kelly. “Receding horizon model-predictive control for mobile robot navigation of intricate paths.” In: *Field and Service Robotics. Results of the 7th International Conference.* Ed. by A. Howard, K. Iagnemma, and A. Kelly. Springer Tracts in Advanced Robotics 62. Berlin: Springer, 2010, pp. 69–78.

[How14] T. M. Howard, M. Pivtoraiko, R. A. Knepper, and A. Kelly. “Model-predictive motion planning.” In: *IEEE Robotics & Automation Magazine* 21.1 (2014), pp. 64–73.

[Hsu02] D. Hsu, R. Kindel, J.-C. Latombe, and S. M. Rock. “Randomized kinodynamic motion planning with moving obstacles.” In: *The International Journal of Robotics Research* 21.3 (2002), pp. 233–256.

[Hsu98] D. Hsu, L. E. Kavraki, J.-C. Latombe, R. Motwani, and S. Sorkin. “On finding narrow passages with probabilistic roadmap planners.” In: *Proceedings of the third Workshop on the Algorithmic Foundations of Robotics: The Algorithmic Perspective.* 1998, pp. 141–153.

[Hug14] J. F. Hughes, A. van Dam, M. McGuire, D. F. Sklar, J. D. Foley, S. K. Feiner, and K. Akeley. *Computer graphics: principles and practice.* 3rd ed. Upper Saddle River, NJ: Addison-Wesley, 2014.

[Iag06] K. Iagnemma and M. Buehler, eds. *Journal of Field Robotics* 23.8–9 (2006): *Special Issue on the DARPA Grand Challenge, Part 1 and 2.*

[Jai04] L. Jaillet and T. Simeon. “A PRM-based motion planner for dynamically changing environments.” In: *Proceedings of the IEEE/RSJ International Conference on Intelligent Robots and Systems.* 2004, pp. 1606–1611.

[Jam08] D. James and J. B. Maitin-Shepard. *Boost.Unordered.* 2008. URL: `http://www.boost.org/doc/libs/1_60_0/doc/html/heap.html` (retrieved on Feb. 4, 2016).

[Jia92] K. Jiang, L. D. Seneviratne, S. W. E. Earies, and W. S. Ko. “Minimum-time smooth path planning for a mobile robot with kinematic constraints.” In: *Proceedings of the IEEE International Workshop on Emerging Technologies and Factory Automation.* 1992, pp. 531–536.

[Kam08] S. Kammel, J. Ziegler, B. Pitzer, M. Werling, T. Gindele, D. Jagzent, J. Schröder, M. Thuy, M. Goebl, F. von Hundelshausen, O. Pink, C. Frese, and C. Stiller. “Team AnnieWAY’s autonomous system for the 2007 DARPA Urban Challenge.” In: *Journal of Field Robotics* 25.9 (2008), pp. 615–639.

[Kan86] K. Kant and S. W. Zucker. “Toward efficient trajectory planning: the path-velocity decomposition.” In: *The International Journal of Robotics Research* 5.3 (1986), pp. 72–89.

[Kar10] S. Karaman and E. Frazzoli. “Optimal kinodynamic motion planning using incremental sampling-based methods.” In: *Proceedings of the 49th IEEE Conference on Decision and Control.* 2010, pp. 7681–7687.

[Kar11] S. Karaman and E. Frazzoli. “Sampling-based algorithms for optimal motion planning.” In: *The International Journal of Robotics Research* 30.7 (2011), pp. 846–894.

[Kar13] S. Karaman and E. Frazzoli. “Sampling-based optimal motion planning for non-holonomic dynamical systems.” In: *Proceedings of the IEEE International Conference on Robotics and Automation.* 2013, pp. 5041–5047.

[Kav96] L. E. Kavraki, P. Švestka, J.-C. Latombe, and M. H. Overmars. "Probabilistic roadmaps for path planning in high-dimensional configuration spaces." In: *IEEE Transactions on Robotics and Automation* 12.4 (1996), pp. 566–580.

[Kel03] A. Kelly and B. Nagy. "Reactive nonholonomic trajectory generation via parametric optimal control." In: *The International Journal of Robotics Research* 22.7-8 (2003), pp. 583–601.

[Kel06] A. Kelly, A. Stentz, O. Amidi, M. Bode, D. Bradley, A. Diaz-Calderon, M. Happold, H. Herman, R. Mandelbaum, T. Pilarski, P. Rander, S. Thayer, N. Vallidis, and R. Warner. "Toward reliable off road autonomous vehicles operating in challenging environments." In: *The International Journal of Robotics Research* 25.5–6 (2006), pp. 449–483.

[Kha86] O. Khatib. "Real-time obstacle avoidance for manipulators and mobile robots." In: *The International Journal of Robotics Research* 5.1 (1986), pp. 90–98.

[Kin09] J. King and M. Likhachev. "Efficient cost computation in cost map planning for non-circular robots." In: *Proceedings of the IEEE/RSJ International Conference on Intelligent Robots and Systems.* 2009, pp. 3924–3930.

[Kle12] L. A. Klein. *Sensor and data fusion. A tool for information assessment and decision making.* 2nd ed. Bellingham, WA: SPIE – The International Society for Optical Engineering, 2012.

[Kne06] R. A. Knepper and A. Kelly. "High performance state lattice planning using heuristic look-up tables." In: *Proceedings of the IEEE/RSJ International Conference on Intelligent Robots and Systems.* 2006, pp. 3375–3380.

[Kne09] R. A. Knepper and M. T. Mason. "Empirical sampling of path sets for local area motion planning." In: *Experimental Robotics. The Eleventh International Symposium.* Ed. by O. Khatib, V. Kumar, and G. Pappas. Springer Tracts in Advanced Robotics 54. Berlin: Springer, 2009, pp. 451–462.

[Kne10] R. A. Knepper, S. S. Srinivasa, and M. T. Mason. "Hierarchical planning architectures for mobile manipulation tasks in indoor environments." In: *Proceedings of the IEEE International Conference on Robotics and Automation.* 2010, pp. 1985–1990.

[Koe05] S. Koenig and M. Likhachev. "Fast replanning for navigation in unknown terrain." In: *IEEE Transactions on Robotics* 21.3 (2005), pp. 354–363.

[Koe06] S. Koenig and M. Likhachev. "A new principle for incremental heuristic search: Theoretical results." In: *Proceedings of the Sixteenth International Conference on Automated Planning and Scheduling.* 2006, pp. 402–405.

[Krü10] P. Krüsi, M. Pivtoraiko, A. Kelly, T. M. Howard, and R. Siegwart. "Path set relaxation for mobile robot navigation." In: *10th International Symposium on Artificial Intelligence, Robotics and Automation in Space.* 2010, pp. 456–463.

[Kuf00] J. J. Kuffner and S. M. LaValle. "RRT-connect: An efficient approach to single-query path planning." In: *Proceedings of the IEEE International Conference on Robotics and Automation.* 2000, pp. 995–1001.

[Kun14] H.-B. Kuntze, C. Frey, T. Emter, J. Petereit, I. Tchouchenkov, T. Müller, M. Tittel, R. Worst, K. Pfeiffer, M. Walter, S. Rademacher, and F. Müller. "Situation responsive networking of mobile robots for disaster management." In: *Proceedings of the 45th International Symposium on Robotics / 8th German Conference on Robotics (ISR/ROBOTIK 2014).* 2014.

[Kus09] A. Kushleyev and M. Likhachev. "Time-bounded lattice for efficient planning in dynamic environments." In: *Proceedings of the IEEE International Conference on Robotics and Automation.* 2009, pp. 1662–1668.

[Kuw08] Y. Kuwata, J. Teo, S. Karaman, G. Fiore, E. Frazzoli, and J. How. "Motion planning in complex environments using closed-loop prediction." In: *AIAA Guidance, Navigation and Control Conference and Exhibit.* 2008.

[Kuw09] Y. Kuwata, J. Teo, G. Fiore, S. Karaman, E. Frazzoli, and J. P. How. "Real-time motion planning with applications to autonomous urban driving." In: *IEEE Transactions on Control Systems Technology* 17.5 (2009), pp. 1105–1118.

[Lac98] A. Lacaze, Y. Moscovitz, N. DeClaris, and K. Murphy. "Path planning for autonomous vehicles driving over rough terrain." In: *Proceedings of the joint IEEE International Symposium on Intelligent Control / Computational Intelligence in Robotics and Automation / Intelligent Systems and Semiotics.* 1998, pp. 50–55.

[Lam01] F. Lamiraux and J.-P. Laumond. "Smooth Motion Planning for Car-Like Vehicles." In: *IEEE Transactions on Robotics and Automation* 17.4 (2001), pp. 498–502.

[Lam08] A. Lambert, D. Gruyer, G. Saint Pierre, and A. N. Ndjeng. "Collision probability assessment for speed control." In: *Proceedings of the 11th International IEEE Conference on Intelligent Transportation Systems.* 2008, pp. 1043–1048.

[Lar02] F. Large, S. Sekhavat, Z. Shiller, and C. Laugier. "Towards real-time global motion planning in a dynamic environment using the NLVO concept." In: *Proceedings of the IEEE/RSJ International Conference on Intelligent Robots and Systems.* 2002, pp. 607–612.

[Lat91] J.-C. Latombe. *Robot Motion Planning.* Dordrecht: Kluwer Academic, 1991.

[Lau87] J.-P. Laumond. "Feasible trajectories for mobile robots with kinematic and environment constraints." In: *Intelligent Autonomous Systems.* Ed. by L. O. Hertzberger and F. C. A. Groen. Amsterdam: North-Holland Publishing Co., 1987, pp. 346–354.

[Lau94] J.-P. Laumond, P. E. Jacobs, M. Taix, and R. M. Murray. "A motion planner for nonholonomic mobile robots." In: *IEEE Transactions on Robotics and Automation* 10.5 (1994), pp. 577–593.

[Lau98] J.-P. Laumond, ed. *Robot Motion Planning and Control.* Lecture Notes in Control and Information Sciences. Berlin: Springer, 1998.

[LaV00] S. M. LaValle and J. J. Kuffner Jr. "Rapidly-exploring random trees: progress and prospects." In: *Algorithmic and Computational Robotics: New Directions.* 2000.

[LaV01] S. M. LaValle and J. J. Kuffner Jr. "Randomized kinodynamic planning." In: *The International Journal of Robotics Research* 20.5 (2001), pp. 378–400.

[LaV04] S. M. LaValle, M. S. Branicky, and S. R. Lindemann. "On the relationship between classical grid search and probabilistic roadmaps." In: *The International Journal of Robotics Research* 23.7–8 (2004), pp. 673–692.

[LaV06] S. M. LaValle. *Planning Algorithms.* Cambridge: Cambridge University Press, 2006.

[LaV11a] S. M. LaValle. "Motion planning: the essentials." In: *IEEE Robotics & Automation Magazine* 18.1 (2011), pp. 79–89.

[LaV11b] S. M. LaValle. "Motion planning: wild frontiers." In: *IEEE Robotics & Automation Magazine* 18.2 (2011), pp. 108–118.

[LaV98] S. M. LaValle. *Rapidly-exploring random trees: a new tool for path planning.* Tech. rep. TR 98-11. Department of Computer Science, Iowa State University, Oct. 1998.

[Lev11] J. Levinson, J. Askeland, J. Becker, J. Dolson, D. Held, S. Kammel, J. Z. Kolter, D. Langer, O. Pink, V. Pratt, M. Sokolsky, G. Stanek, D. Stavens, A. Teichman, M. Werling, and S. Thrun. "Towards fully autonomous driving: Systems and algorithms." In: *Proceedings of the IEEE Intelligent Vehicles Symposium.* 2011, pp. 163–168.

[Lik03a] M. Likhachev, G. Gordon, and S. Thrun. "ARA*: anytime A* with provable bounds on sub-optimality." In: *Proceedings of the Conference on Neural Information Processing Systems.* 2003.

[Lik03b] M. Likhachev, G. Gordon, and S. Thrun. *ARA*: formal analysis.* Tech. rep. CMU-CS-03-148. School of Computer Science, Carnegie Mellon University, July 2003.

[Lik05] M. Likhachev, D. Ferguson, G. Gordon, A. Stentz, and S. Thrun. "Anytime Dynamic A*: an anytime, replanning algorithm." In: *Proceedings of the International Conference on Automated Planning and Scheduling.* 2005, pp. 262–271.

[Lik08] M. Likhachev, D. Ferguson, G. Gordon, A. Stentz, and S. Thrun. "Anytime search in dynamic graphs." In: *Artificial Intelligence* 172.14 (2008), pp. 1613–1643.

[Lik09] M. Likhachev and D. Ferguson. "Planning long dynamically feasible maneuvers for autonomous vehicles." In: *The International Journal of Robotics Research* 28.8 (2009), pp. 933–945.

[Lik16] M. Likhachev et al. *Search-Based Planning Lab.* URL: `http://sbpl.net` (retrieved on Feb. 4, 2016).

[Lin03] S. R. Lindemann and S. M. LaValle. "Incremental low-discrepancy lattice methods for motion planning." In: *Proceedings of the IEEE International Conference on Robotics and Automation.* 2003, pp. 2920–2927.

[Lin04] S. R. Lindemann and S. M. LaValle. "Steps toward derandomizing RRTs." In: *Proceedings of the Fourth International Workshop on Robot Motion and Control.* 2004, pp. 271–277.

[Lin05] S. R. Lindemann and S. M. LaValle. "Smoothly blending vector fields for global robot navigation." In: *44th IEEE Conference on Decision and Control, and European Control Conference.* 2005, pp. 3553–3559.

[Lin06] S. R. Lindemann and S. M. LaValle. "A multiresolution approach for motion planning under differential constraints." In: *Proceedings of the IEEE International Conference on Robotics and Automation.* 2006, pp. 139–144.

[Lin07] S. R. Lindemann and S. M. LaValle. "Smooth feedback for car-like vehicles in polygonal environments." In: *Proceedings of the IEEE International Conference on Robotics and Automation.* 2007, pp. 3104–3109.

[Lin08] R. Linker and T. Blass. "Optimal path planning for car-like off-road vehicles." In: *Proceedings of the IEEE Conference on Robotics, Automation and Mechatronics.* 2008, pp. 150–154.

[Lin09] S. R. Lindemann and S. M. LaValle. "Simple and efficient algorithms for computing smooth, collision-free feedback laws over given cell decompositions." In: *The International Journal of Robotics Research* 28.5 (2009), pp. 600–621.

[Loz79] T. Lozano-Pérez and M. A. Wesley. "An algorithm for planning collision-free paths among polyhedral obstacles." In: *Commununications of the ACM* 22.10 (1979), pp. 560–570.

[Mag07] M. Magnusson, A. Lilienthal, and T. Duckett. "Scan registration for autonomous mining vehicles using 3D-NDT." In: *Journal of Field Robotics* 24.10 (2007), pp. 803–827.

[Mau03] C. R. Maurer Jr., R. Qi, and V. Raghavan. "A linear time algorithm for computing exact Euclidean distance transforms of binary images in arbitrary dimensions." In: *IEEE Transactions on Pattern Analysis and Machine Intelligence* 25.2 (2003), pp. 265–270.

[McN11] M. McNaughton, C. Urmson, J. M. Dolan, and J.-W. Lee. "Motion planning for autonomous driving with a conformal spatiotemporal lattice." In: *Proceedings of the IEEE International Conference on Robotics and Automation.* 2011, pp. 4889–4895.

[Mil10] G. Milighetti, J. Petereit, and H.-B. Kuntze. "Mobile experimental platform for the development of environmentally interactive control algorithms towards the implementation on a walking humanoid robot." In: *Proceedings of the 41st International Symposium on Robotics/6th German Conference on Robotics (ISR/ROBOTIK 2010).* 2010.

[Mon07] M. Montemerlo and S. Thrun. *FastSLAM. A Scalable Method for the Simultaneous Localization and Mapping Problem in Robotics.* Springer Tracts in Advanced Robotics 27. Berlin: Springer, 2007.

[Mon08] M. Montemerlo, J. Becker, S. Bhat, H. Dahlkamp, D. Dolgov, S. Ettinger, D. Haehnel, T. Hilden, G. Hoffmann, B. Huhnke, D. Johnston, S. Klumpp, D. Langer, A. Levandowski, J. Levinson, J. Marcil, D. Orenstein, J. Paefgen, I. Penny, A. Petrovskaya, M. Pflueger, G. Stanek, D. Stavens, A. Vogt, and S. Thrun. "Junior: The Stanford entry in the

Urban Challenge." In: *Journal of Field Robotics* 25.9 (2008), pp. 569–597.

[Mor04] S. Morgan and M. S. Branicky. "Sampling-based planning for discrete spaces." In: *Proceedings of the IEEE/RSJ International Conference on Intelligent Robots and Systems.* 2004, pp. 1938–1945.

[Mou06] A. Mourikis and S. Roumeliotis. "On the treatment of relative-pose measurements for mobile robot localization." In: *Proceedings of the IEEE International Conference on Robotics and Automation.* 2006, pp. 2277–2284.

[Muñ10] N. D. Muñoz Ceballos, J. A. Valencia, and N. L. Ospina. "Quantitative performance metrics for mobile robots navigation." In: *Mobile Robots Navigation.* Ed. by A. Barrera. Rijeka: InTech, 2010.

[Mur04] R. R. Murphy. "Human-robot interaction in rescue robotics." In: *IEEE Transactions on Systems, Man, and Cybernetics, Part C (Applications and Reviews)* 34.2 (2004), pp. 138–153.

[Nag13] K. Nagatani, S. Kiribayashi, Y. Okada, K. Otake, K. Yoshida, S. Tadokoro, T. Nishimura, T. Yoshida, E. Koyanagi, M. Fukushima, and S. Kawatsuma. "Emergency response to the nuclear accident at the Fukushima Daiichi Nuclear Power Plants using mobile rescue robots." In: *Journal of Field Robotics* 30.1 (2013), pp. 44–63.

[Neu09] F. Neuhaus, D. Dillenberger, J. Pellenz, and D. Paulus. "Terrain drivability analysis in 3D laser range data for autonomous robot navigation in unstructured environments." In: *Proceedings of the IEEE Conference on Emerging Technologies & Factory Automation.* 2009.

[Now10] W. Nowak, A. Zakharov, S. Blumenthal, and E. Prassler. *Benchmarks for mobile manipulation and robust obstacle avoidance and navigation.* Deliverable D3.1. Best Practice in Robotics (BRICS), FP7 grant agreement no. 231940, Apr. 2010.

[ÓDú85] C. Ó'Dúnlaing and C. K. Yap. "A 'retraction' method for planning the motion of a disc." In: *Journal of Algorithms* 6.1 (1985), pp. 104–111.

[Olv10] F. W. J. Olver, D. W. Lozier, R. F. Boisvert, and C. W. Clark, eds. *NIST Handbook of Mathematical Functions.* Cambridge: Cambridge University Press, 2010.

[Pan04] S. Pancanti, L. Pallottino, D. Salvadorini, and A. Bicchi. "Motion planning through symbols and lattices." In: *Proceedings of the IEEE International Conference on Robotics and Automation.* 2004, pp. 3914–3919.

[Pea84] J. Pearl. *Heuristics. Intelligent search strategies for computer problem solving.* Artificial Intelligence. Reprinted with corrections October, 1985. Reading, MA: Addison-Wesley, 1984.

[Pet10] J. Petereit and T. Bernard. "Real-time nonlinear model predictive control of a glass forming process using a finite element model." In: *Proceedings of the 8th IFAC Symposium on Nonlinear Control Systems.* 2010.

[Pet12] J. Petereit, T. Emter, C. W. Frey, T. Kopfstedt, and A. Beutel. "Application of Hybrid A* to an autonomous mobile robot for path planning in unstructured outdoor environments." In: *Proceedings of the 7th German Conference on Robotics (ROBOTIK 2012).* 2012.

[Pet13a] J. Petereit and T. Emter. "Kombinierte Pfad- und Trajektorienplanung für autonome mobile Roboter." In: *Tagungsband des 19. Workshop Computer-Bildanalyse in der Landwirtschaft und 2. Workshop Unbemannte autonom fliegende Systeme (UAS) in der Landwirtschaft.* Bornimer Agrartechnische Berichte 81. Leibniz-Institut für Agrartechnik Potsdam-Bornim e. V., 2013, pp. 24–30.

[Pet13b] J. Petereit, T. Emter, and C. W. Frey. "Mobile robot motion planning in multi-resolution lattices with hybrid dimensionality." In: *Proceedings of the IFAC Intelligent Autonomous Vehicles Symposium.* 2013.

[Pet13c] J. Petereit, T. Emter, and C. W. Frey. "Safe mobile robot motion planning for waypoint sequences in a dynamic environment." In: *Proceedings of the IEEE International Conference on Information Technology.* 2013.

[Pet14] J. Petereit, T. Emter, and C. W. Frey. “Combined trajectory generation and path planning for mobile robots using lattices with hybrid dimensionality.” In: *Robot Intelligence Technology and Applications 2. Results from the 2nd International Conference on Robot Intelligence Technology and Applications.* Ed. by J.-H. Kim, E. T. Matson, H. Myung, P. Xu, and F. Karray. Advances in Intelligent Systems and Computing 274. Springer International Publishing Switzerland, 2014, pp. 145–157.

[Phi03] R. Philippsen and R. Siegwart. “Smooth and efficient obstacle avoidance for a tour guide robot.” In: *Proceedings of the IEEE International Conference on Robotics and Automation.* 2003, pp. 446–451.

[Phi07] R. Philippsen, S. Kolski, K. Maček, and R. Siegwart. “Path planning, replanning, and execution for autonomous driving in urban and offroad environments.” In: *IEEE International Conference on Robotics and Automation: Workshop on Planning, Perception and Navigation for Intelligent Vehicles.* 2007.

[Phi11] M. Phillips and M. Likhachev. “SIPP: Safe interval path planning for dynamic environments.” In: *Proceedings of the IEEE International Conference on Robotics and Automation.* 2011, pp. 5628–5635.

[Phi14] M. Phillips, M. Likhachev, and S. Koenig. “PA*SE: parallel A* for slow expansions.” In: *Proceedings of the International Conference on Automated Planning and Scheduling.* 2014.

[Piv04] M. Pivtoraiko and A. Kelly. *A study of polynomial curvature clothoid paths for motion planning for car-like robots.* Tech. rep. CMU-RI-TR-04-68. Robotics Institute, Carnegie Mellon University, Dec. 2004.

[Piv05a] M. Pivtoraiko and A. Kelly. “Generating near minimal spanning control sets for constrained motion planning in discrete state spaces.” In: *Proceedings of the IEEE/RSJ International Conference on Intelligent Robots and Systems.* 2005, pp. 3231–3237.

[Piv05b] M. Pivtoraiko and A. Kelly. “Efficient constrained path planning via search in state lattices.” In: *Proceedings of the 8th International Symposium on Artificial Intelligence, Robotics and Automation in Space.* 2005.

[Piv06] M. Pivtoraiko and A. Kelly. "Constrained motion planning in discrete state spaces." In: *Field and Service Robotics. Results of the 5th International Conference.* Ed. by P. Corke and S. Sukkarieh. Springer Tracts in Advanced Robotics 25. Berlin: Springer, 2006, pp. 269–280.

[Piv07] M. Pivtoraiko, R. A. Knepper, and A. Kelly. *Optimal, smooth, nonholonomic mobile robot motion planning in state lattices.* Tech. rep. CMU-RI-TR-07-15. Robotics Institute, Carnegie Mellon University, May 2007.

[Piv08] M. Pivtoraiko and A. Kelly. "Differentially constrained motion replanning using state lattices with graduated fidelity." In: *Proceedings of the IEEE/RSJ International Conference on Intelligent Robots and Systems.* 2008, pp. 2611–2616.

[Piv09a] M. Pivtoraiko. "Adaptive anytime motion planning for robust robot navigation in natural environments." In: *Advanced Technologies for Enhanced Quality of Life.* 2009, pp. 123–129.

[Piv09b] M. Pivtoraiko and A. Kelly. "Fast and feasible deliberative motion planner for dynamic environments." In: *IEEE International Conference on Robotics and Automation: Workshop on Safe Navigation in Open and Dynamic Environments: Application to Autonomous Vehicles.* 2009.

[Piv09c] M. Pivtoraiko, R. A. Knepper, and A. Kelly. "Differentially constrained mobile robot motion planning in state lattices." In: *Journal of Field Robotics* 26.3 (2009), pp. 308–333.

[Piv11] M. Pivtoraiko and A. Kelly. "Kinodynamic motion planning with state lattice motion primitives." In: *Proceedings of the IEEE/RSJ International Conference on Intelligent Robots and Systems.* 2011, pp. 2172–2179.

[Piv12] M. Pivtoraiko. "Differentially constrained motion planning with state lattice motion primitives." PhD thesis. Robotics Institute, Carnegie Mellon University, Feb. 2012.

[Poh70] I. Pohl. "Heuristic search viewed as path finding in a graph." In: *Artificial Intelligence* 1.3 (1970), pp. 193–204.

[Qui09] M. Quigley, K. Conley, B. P. Gerkey, J. Faust, T. Foote, J. Leibs, R. Wheeler, and A. Y. Ng. "ROS: an open-source robot operating system." In: *IEEE International Conference on Robotics and Automation: Workshop on Open Source Software.* 2009.

[Qui93] S. Quinlan and O. Khatib. "Elastic bands: connecting path planning and control." In: *Proceedings of the IEEE International Conference on Robotics and Automation.* 1993, pp. 802–807.

[Ree90] J. A. Reeds and R. A. Shepp. "Optimal paths for a car that goes both forwards and backwards." In: *Pacific Journal of Mathematics* 145.2 (1990), pp. 367–393.

[Reu98] J. Reuter. "Mobile robots trajectories with continuously differentiable curvature: an optimal control approach." In: *Proceedings of the IEEE/RSJ International Conference on Intelligent Robots and Systems.* 1998, pp. 38–43.

[ROS16] *ROS.org | Powering the world's robots.* URL: `http://www.ros.org` (retrieved on Feb. 4, 2016).

[Ruf09a] M. Rufli, D. Ferguson, and R. Siegwart. "Smooth path planning in constrained environments." In: *Proceedings of the IEEE International Conference on Robotics and Automation.* 2009, pp. 3780–3785.

[Ruf09b] M. Rufli and R. Siegwart. "On the application of the D* search algorithm to time-based planning on lattice graphs." In: *Proceedings of The 4th European Conference on Mobile Robotics.* 2009.

[Ruf10] M. Rufli and R. Siegwart. "On the design of deformable input-/state-lattice graphs." In: *Proceedings of the IEEE International Conference on Robotics and Automation.* 2010, pp. 3071–3077.

[Rus95] S. J. Russell and P. Norvig. *Artificial Intelligence. A Modern Approach.* Englewood Cliffs, NJ: Prentice-Hall, 1995.

[San01] G. Sanchéz and J.-C. Latombe. "A single-query bi-directional probabilistic roadmap planner with lazy collision checking." In: *Proceedings of the International Symposium on Robotics Research.* 2001.

[Sch98] A. Scheuer and C. Laugier. "Planning sub-optimal and continuous-curvature paths for car-like robots." In: *Proceedings of the IEEE/RSJ International Conference on Intelligent Robots and Systems.* 1998, pp. 25–31.

[Sek98] S. Sekhavat, P. Švestka, J.-P. Laumond, and M. H. Overmars. "Multi-level path planning for nonholonomic robots using semiholonomic subsystems." In: *The International Journal of Robotics Research* 17.8 (1998), pp. 840–857.

[Sek99] S. Sekhavat and M. Chyba. "Nonholonomic deformation of a potential field for motion planning." In: *Proceedings of the IEEE International Conference on Robotics and Automation.* 1999, pp. 817–822.

[Shi07] Z. Shiller, F. Large, S. Sekhavat, and C. Laugier. "Motion planning in dynamic environments." In: *Autonomous Navigation in Dynamic Environments.* Ed. by C. Laugier and R. Chatila. Springer Tracts in Advanced Robotics 35. Berlin: Springer, 2007, pp. 107–119.

[Shi91] Z. Shiller and Y.-R. Gwo. "Dynamic motion planning of autonomous vehicles." In: *IEEE Transactions on Robotics and Automation* 7.2 (1991), pp. 241–249.

[Sim96] R. Simmons. "The curvature-velocity method for local obstacle avoidance." In: *Proceedings of the IEEE International Conference on Robotics and Automation.* 1996, pp. 3375–3382.

[Ste94] A. Stentz. "Optimal and efficient path planning for partially-known environments." In: *Proceedings of the IEEE International Conference on Robotics and Automation.* 1994, pp. 3310–3317.

[Stu12] N. R. Sturtevant, A. Felner, M. Likhachev, and W. Ruml. "Heuristic search comes of age." In: *Proceedings of the AAAI Conference on Artificial Intelligence.* 2012, pp. 2186–2191.

[Şuc12a] I. A. Şucan and L. E. Kavraki. "A sampling-based tree planner for systems with complex dynamics." In: *IEEE Transactions on Robotics* 28.1 (2012), pp. 116–131.

[Şuc12b] I. A. Şucan, M. Moll, and L. E. Kavraki. "The Open Motion Planning Library." In: *IEEE Robotics & Automation Magazine* 19.4 (2012), pp. 72–82. URL: `http://ompl.kavrakilab.org`.

[Sun05] Z. Sun, D. Hsu, T. Jiang, H. Kurniawati, and J. H. Reif. "Narrow passage sampling for probabilistic roadmap planning." In: *IEEE Transactions on Robotics* 21.6 (2005), pp. 1105–1115.

[Šve97] P. Švestka and M. H. Overmars. "Motion planning for carlike robots using a probabilistic learning approach." In: *The International Journal of Robotics Research* 16.2 (1997), pp. 119–143.

[Thr06a] S. Thrun, W. Burgard, and D. Fox. *Probabilistic Robotics.* Cambridge, MA: MIT Press, 2006.

[Thr06b] S. Thrun, M. Montemerlo, H. Dahlkamp, D. Stavens, A. Aron, J. Diebel, P. Fong, J. Gale, M. Halpenny, G. Hoffmann, K. Lau, C. Oakley, M. Palatucci, V. Pratt, P. Stang, S. Strohband, C. Dupont, L.-E. Jendrossek, C. Koelen, C. Markey, C. Rummel, J. van Niekerk, E. Jensen, P. Alessandrini, G. Bradski, B. Davies, S. Ettinger, A. Kaehler, A. Nefian, and P. Mahoney. "Stanley: The robot that won the DARPA Grand Challenge." In: *Journal of Field Robotics* 23.9 (2006), pp. 661–692.

[Thr96] S. Thrun and A. Bücken. "Integrating grid-based and topological maps for mobile robot navigation." In: *Proceedings of the Thirteenth National Conference on Artificial Intelligence AAAI.* 1996.

[Urm06] C. Urmson, C. Ragusa, D. Ray, J. Anhalt, D. Bartz, T. Galatali, A. Gutierrez, J. Johnston, S. Harbaugh, H. "Yu" Kato, W. Messner, N. Miller, K. Peterson, B. Smith, J. Snider, S. Spiker, J. Ziglar, W. "Red" Whittaker, M. Clark, P. Koon, A. Mosher, and J. Struble. "A robust approach to high-speed navigation for unrehearsed desert terrain." In: *Journal of Field Robotics* 23.8 (2006), pp. 467–508.

[Urm08] C. Urmson, J. Anhalt, D. Bagnell, C. Baker, R. Bittner, M. N. Clark, J. Dolan, D. Duggins, T. Galatali, C. Geyer, M. Gittleman, S. Harbaugh, M. Hebert, T. M. Howard, S. Kolski, A. Kelly, M. Likhachev, M. McNaughton, N. Miller, K. Peterson, B. Pilnick, R. Rajkumar, P. Rybski, B.

Salesky, Y.-W. Seo, S. Singh, J. Snider, A. Stentz, W. "Red" Whittaker, Z. Wolkowicki, J. Ziglar, H. Bae, T. Brown, D. Demitrish, B. Litkouhi, J. Nickolaou, V. Sadekar, W. Zhang, J. Struble, M. Taylor, M. Darms, and D. Ferguson. "Autonomous driving in urban environments: Boss and the Urban Challenge." In: *Journal of Field Robotics* 25.8 (2008), pp. 425–466.

[Vah08] N. Vahrenkamp, C. Scheurer, T. Asfour, J. Kuffner, and R. Dillmann. "Adaptive motion planning for humanoid robots." In: *Proceedings of the IEEE/RSJ International Conference on Intelligent Robots and Systems.* 2008, pp. 2127–2132.

[Wer10] M. Werling, J. Ziegler, S. Kammel, and S. Thrun. "Optimal trajectory generation for dynamic street scenarios in a Frénet frame." In: *Proceedings of the IEEE International Conference on Robotics and Automation.* 2010, pp. 987–993.

[Wer11] M. Werling, S. Kammel, J. Ziegler, and L. Gröll. "Optimal trajectories for time-critical street scenarios using discretized terminal manifolds." In: *The International Journal of Robotics Research* 31.3 (2011), pp. 346–359.

[Xu12] W. Xu, J. Wei, J. M. Dolan, H. Zhao, and H. Zha. "A real-time motion planner with trajectory optimization for autonomous vehicles." In: *Proceedings of the IEEE International Conference on Robotics and Automation.* 2012, pp. 2061–2067.

[Zha12] H. Zhang, J. Butzke, and M. Likhachev. "Combining global and local planning with guarantees on completeness." In: *Proceedings of the IEEE International Conference on Robotics and Automation.* 2012, pp. 4500–4506.

[Zho15] Y. Zhou and J. Zeng. "Massively parallel A* search on a GPU." In: *Proceedings of the Twenty-Ninth AAAI Conference on Artificial Intelligence.* 2015, pp. 1248–1254.

[Zie08] J. Ziegler, M. Werling, and J. Schröder. "Navigating car-like robots in unstructured environments using an obstacle sensitive cost function." In: *Proceedings of the IEEE Intelligent Vehicles Symposium.* 2008, pp. 787–791.

[Zie09] J. Ziegler and C. Stiller. "Spatiotemporal state lattices for fast trajectory planning in dynamic on-road driving scenarios." In: *Proceedings of the IEEE/RSJ International Conference on Intelligent Robots and Systems.* 2009, pp. 1879–1884.

[Zie10] J. Ziegler and C. Stiller. "Fast collision checking for intelligent vehicle motion planning." In: *Proceedings of the IEEE Intelligent Vehicles Symposium.* 2010, pp. 518–522.

[Zie14] J. Ziegler, P. Bender, M. Schreiber, H. Lategahn, T. Strauss, C. Stiller, T. Dang, U. Franke, N. Appenrodt, C. G. Keller, E. Kaus, R. G. Herrtwich, C. Rabe, D. Pfeiffer, F. Lindner, F. Stein, F. Erbs, M. Enzweiler, C. Knöppel, J. Hipp, M. Haueis, M. Trepte, C. Brenk, A. Tamke, M. Ghanaat, M. Braun, A. Joos, H. Fritz, H. Mock, M. Hein, and E. Zeeb. "Making Bertha drive – An autonomous journey on a historic route." In: *IEEE Intelligent Transportation Systems Magazine* 6.2 (Summer 2014), pp. 8–20.

[Zie15] J. Ziehn, M. Ruf, B. Rosenhahn, D. Willersinn, J. Beyerer, and H. Gotzig. "Correspondence between variational methods and Hidden Markov Models." In: *Proceedings of the IEEE Intelligent Vehicles Symposium.* 2015, pp. 380–385.

[Zuc07] M. Zucker, J. Kuffner, and M. Branicky. "Multipartite RRTs for rapid replanning in dynamic environments." In: *Proceedings of the IEEE International Conference on Robotics and Automation.* 2007, pp. 1603–1609.

List of Publications

[Cur15] F. Curatella, P. Vinetti, G. Rizzo, T. Vladimirova, L. D. Vendictis, T. Emter, J. Petereit, C. W. Frey, D. Usher, I. Stanciugelu, J. Schaefer, E. den Breejen, L. Gisslén, and D. Letalick. "Toward a multifaceted platform for humanitarian demining." In: *13th IARP Workshop on Humanitarian Demining and Risky Intervention.* 12th International Symposium "MINE ACTION 2015". 2015, pp. 133–144.

[Fre15] C. Frey, T. Emter, J. Petereit, I. Tchouchenkov, T. Müller, M. Tittel, R. Worst, K. Pfeiffer, M. Walter, J. Wöllenstein, S. Rademacher, A. Wenzel, and F. Müller. "Situation responsive networking of mobile robots for disaster management." In: *Proceedings of the IEEE International Symposium on Technologies for Homeland Security.* 2015.

[Emt14] T. Emter and J. Petereit. "Integrated multi-sensor fusion for mapping and localization in outdoor environments for mobile robots." In: *Proceedings of SPIE, volume 9121. Multisensor, Multisource Information Fusion: Architectures, Algorithms, and Applications.* 2014.

[Kun14] H.-B. Kuntze, C. Frey, T. Emter, J. Petereit, I. Tchouchenkov, T. Müller, M. Tittel, R. Worst, K. Pfeiffer, M. Walter, S. Rademacher, and F. Müller. "Situation responsive networking of mobile robots for disaster management." In: *Proceedings of the 45th International Symposium on Robotics / 8th German Conference on Robotics (ISR/ROBOTIK 2014).* 2014.

[Pet14] J. Petereit, T. Emter, and C. W. Frey. "Combined trajectory generation and path planning for mobile robots using lattices with hybrid dimensionality." In: *Robot Intelligence Technology and Applications 2. Results from the 2nd International Conference on Robot Intelligence Technology and Applications.* Ed. by J.-H. Kim, E. T. Matson, H. Myung, P. Xu, and F. Karray. Advances in Intelligent Systems and Computing 274. Springer International Publishing Switzerland, 2014, pp. 145–157.

[Pet13a] J. Petereit and T. Emter. "Kombinierte Pfad- und Trajektorienplanung für autonome mobile Roboter." In: *Tagungsband des 19. Workshop Computer-Bildanalyse in der Landwirtschaft und 2. Workshop Unbemannte autonom fliegende Systeme (UAS) in der Landwirtschaft.* Bornimer Agrartechnische Berichte 81. Leibniz-Institut für Agrartechnik Potsdam-Bornim e. V., 2013, pp. 24–30.

[Pet13b] J. Petereit, T. Emter, and C. W. Frey. "Mobile robot motion planning in multi-resolution lattices with hybrid dimensionality." In: *Proceedings of the IFAC Intelligent Autonomous Vehicles Symposium.* 2013.

[Pet13c] J. Petereit, T. Emter, and C. W. Frey. "Safe mobile robot motion planning for waypoint sequences in a dynamic environment." In: *Proceedings of the IEEE International Conference on Information Technology.* 2013.

[Pet12] J. Petereit, T. Emter, C. W. Frey, T. Kopfstedt, and A. Beutel. "Application of Hybrid A* to an autonomous mobile robot for path planning in unstructured outdoor environments." In: *Proceedings of the 7th German Conference on Robotics (ROBOTIK 2012).* 2012.

[Emt10] T. Emter, A. Saltoğlu, and J. Petereit. "Multi-sensor fusion for localization of a mobile robot in outdoor environments." In: *Proceedings of the 41st International Symposium on Robotics / 6th German Conference on Robotics (ISR/ROBOTIK 2010).* 2010.

[Mil10] G. Milighetti, J. Petereit, and H.-B. Kuntze. "Mobile experimental platform for the development of environmentally interactive control algorithms towards the implementation on a walking humanoid robot."

In: *Proceedings of the 41st International Symposium on Robotics/6th German Conference on Robotics (ISR/ROBOTIK 2010)*. 2010.

[Pet10] J. Petereit and T. Bernard. "Real-time nonlinear model predictive control of a glass forming process using a finite element model." In: *Proceedings of the 8th IFAC Symposium on Nonlinear Control Systems*. 2010.

Karlsruher Schriftenreihe zur Anthropomatik (ISSN 1863-6489)

Herausgeber: Prof. Dr.-Ing. Jürgen Beyerer

Die Bände sind unter www.ksp.kit.edu als PDF frei verfügbar oder als Druckausgabe bestellbar.

Band 1 Jürgen Geisler
Leistung des Menschen am Bildschirmarbeitsplatz. 2006
ISBN 3-86644-070-7

Band 2 Elisabeth Peinsipp-Byma
Leistungserhöhung durch Assistenz in interaktiven Systemen zur Szenenanalyse. 2007
ISBN 978-3-86644-149-1

Band 3 Jürgen Geisler, Jürgen Beyerer (Hrsg.)
Mensch-Maschine-Systeme. 2010
ISBN 978-3-86644-457-7

Band 4 Jürgen Beyerer, Marco Huber (Hrsg.)
Proceedings of the 2009 Joint Workshop of Fraunhofer IOSB and Institute for Anthropomatics, Vision and Fusion Laboratory. 2010
ISBN 978-3-86644-469-0

Band 5 Thomas Usländer
Service-oriented design of environmental information systems. 2010
ISBN 978-3-86644-499-7

Band 6 Giulio Milighetti
Multisensorielle diskret-kontinuierliche Überwachung und Regelung humanoider Roboter. 2010
ISBN 978-3-86644-568-0

Band 7 Jürgen Beyerer, Marco Huber (Hrsg.)
Proceedings of the 2010 Joint Workshop of Fraunhofer IOSB and Institute for Anthropomatics, Vision and Fusion Laboratory. 2011
ISBN 978-3-86644-609-0

Band 8 Eduardo Monari
Dynamische Sensorselektion zur auftragsorientierten Objektverfolgung in Kameranetzwerken. 2011
ISBN 978-3-86644-729-5

Band 9 Thomas Bader
Multimodale Interaktion in Multi-Display-Umgebungen. 2011
ISBN 3-86644-760-8

Band 10 Christian Frese
Planung kooperativer Fahrmanöver für kognitive Automobile. 2012
ISBN 978-3-86644-798-1

Band 11 Jürgen Beyerer, Alexey Pak (Hrsg.)
Proceedings of the 2011 Joint Workshop of Fraunhofer IOSB and Institute for Anthropomatics, Vision and Fusion Laboratory. 2012
ISBN 978-3-86644-855-1

Band 12 Miriam Schleipen
Adaptivität und Interoperabilität von Manufacturing Execution Systemen (MES). 2013
ISBN 978-3-86644-955-8

Band 13 Jürgen Beyerer, Alexey Pak (Hrsg.)
Proceedings of the 2012 Joint Workshop of Fraunhofer IOSB and Institute for Anthropomatics, Vision and Fusion Laboratory. 2013
ISBN 978-3-86644-988-6

Band 14 Hauke-Hendrik Vagts
Privatheit und Datenschutz in der intelligenten Überwachung: Ein datenschutzgewährendes System, entworfen nach dem „Privacy by Design" Prinzip. 2013
ISBN 978-3-7315-0041-4

Band 15 Christian Kühnert
Data-driven Methods for Fault Localization in Process Technology. 2013
ISBN 978-3-7315-0098-8

Band 16 Alexander Bauer
Probabilistische Szenenmodelle für die Luftbildauswertung. 2014
ISBN 978-3-7315-0167-1

Band 17 Jürgen Beyerer, Alexey Pak (Hrsg.)
Proceedings of the 2013 Joint Workshop of Fraunhofer IOSB and Institute for Anthropomatics, Vision and Fusion Laboratory. 2014
ISBN 978-3-7315-0212-8

Band 18 Michael Teutsch
Moving Object Detection and Segmentation for Remote Aerial Video Surveillance. 2015
ISBN 978-3-7315-0320-0

Band 19 Marco Huber
Nonlinear Gaussian Filtering: Theory, Algorithms, and Applications. 2015
ISBN 978-3-7315-0338-5

Band 20 Jürgen Beyerer, Alexey Pak (Hrsg.)
Proceedings of the 2014 Joint Workshop of Fraunhofer IOSB and Institute for Anthropomatics, Vision and Fusion Laboratory. 2014
ISBN 978-3-7315-0401-6

Band 21 Todor Dimitrov
Permanente Optimierung dynamischer Probleme der Fertigungssteuerung unter Einbeziehung von Benutzerinteraktionen. 2015
ISBN 978-3-7315-0426-9

Band 22 Benjamin Kühn
Interessengetriebene audiovisuelle Szenenexploration. 2016
ISBN 978-3-7315-0457-3

Band 23 Yvonne Fischer
Wissensbasierte probabilistische Modellierung für die Situationsanalyse am Beispiel der maritimen Überwachung. 2016
ISBN 978-3-7315-0460-3

Band 24 Jürgen Beyerer, Alexey Pak (Hrsg.)
Proceedings of the 2015 Joint Workshop of Fraunhofer IOSB and Institute for Anthropomatics, Vision and Fusion Laboratory. 2016
ISBN 978-3-7315-0519-8

Band 25 Pascal Birnstill
Privacy-Respecting Smart Video Surveillance Based on Usage Control Enforcement. 2016
ISBN 978-3-7315-0538-9

Band 26 Philipp Woock
Umgebungskartenschätzung aus Sidescan-Sonardaten für ein autonomes Unterwasserfahrzeug. 2016
ISBN 978-3-7315-0541-9

Band 27 Janko Petereit
Adaptive State × Time Lattices: A Contribution to Mobile Robot Motion Planning in Unstructured Dynamic Environments. 2017
ISBN 978-3-7315-0580-8